AF832677

Research for Development

Series Editors

Emilio Bartezzaghi, Milan, Italy

Giampio Bracchi, Milan, Italy

Adalberto Del Bo, Politecnico di Milano, Milan, Italy

Ferran Sagarra Trias, Department of Urbanism and Regional Planning, Universitat Politècnica de Catalunya, Barcelona, Spain

Francesco Stellacci, Supramolecular NanoMaterials and Interfaces Laboratory (SuNMiL), Institute of Materials, Ecole Polytechnique Fédérale de Lausanne (EPFL), Lausanne, Switzerland

Enrico Zio, Politecnico di Milano, Milan, Italy, Ecole Centrale Paris, Paris, France

The series Research for Development serves as a vehicle for the presentation and dissemination of complex research and multidisciplinary projects. The published work is dedicated to fostering a high degree of innovation and to the sophisticated demonstration of new techniques or methods.

The aim of the Research for Development series is to promote well-balanced sustainable growth. This might take the form of measurable social and economic outcomes, in addition to environmental benefits, or improved efficiency in the use of resources; it might also involve an original mix of intervention schemes.

Research for Development focuses on the following topics and disciplines:

Urban regeneration and infrastructure, Info-mobility, transport, and logistics, Environment and the land, Cultural heritage and landscape, Energy, Innovation in processes and technologies, Applications of chemistry, materials, and nanotechnologies, Material science and biotechnology solutions, Physics results and related applications and aerospace, Ongoing training and continuing education.

Fondazione Politecnico di Milano collaborates as a special co-partner in this series by suggesting themes and evaluating proposals for new volumes. Research for Development addresses researchers, advanced graduate students, and policy and decision-makers around the world in government, industry, and civil society.

THE SERIES IS INDEXED IN SCOPUS

Rossella Moscarelli · Pietro Belotti · Luca Pirolo
Editors

The Future of Slow Mobility

A Review on Design, Planning and Ethical Aspects of E-Bikes

Fondazione Politecnico di Milano

Springer

Editors
Rossella Moscarelli
Department of Architecture and Urban Studies
Politecnico di Milano
Milan, Italy

Pietro Belotti
Department of Electronics, Information, and Bioengineering
Politecnico di Milano
Milan, Italy

Luca Pirolo
Department of Electronics, Information, and Bioengineering
Politecnico di Milano
Milan, Italy

ISSN 2198-7300 ISSN 2198-7319 (electronic)
Research for Development
ISBN 978-3-032-23659-3 ISBN 978-3-032-23660-9 (eBook)
https://doi.org/10.1007/978-3-032-23660-9

© The Editor(s) (if applicable) and The Author(s), under exclusive license to Springer Nature Switzerland AG 2026

This work is subject to copyright. All rights are solely and exclusively licensed by the Publisher, whether the whole or part of the material is concerned, specifically the rights of translation, reprinting, reuse of illustrations, recitation, broadcasting, reproduction on microfilms or in any other physical way, and transmission or information storage and retrieval, electronic adaptation, computer software, or by similar or dissimilar methodology now known or hereafter developed.
The use of general descriptive names, registered names, trademarks, service marks, etc. in this publication does not imply, even in the absence of a specific statement, that such names are exempt from the relevant protective laws and regulations and therefore free for general use.
The publisher, the authors and the editors are safe to assume that the advice and information in this book are believed to be true and accurate at the date of publication. Neither the publisher nor the authors or the editors give a warranty, expressed or implied, with respect to the material contained herein or for any errors or omissions that may have been made. The publisher remains neutral with regard to jurisdictional claims in published maps and institutional affiliations.

This Springer imprint is published by the registered company Springer Nature Switzerland AG
The registered company address is: Gewerbestrasse 11, 6330 Cham, Switzerland

If disposing of this product, please recycle the paper.

Contents

Challenging Slow Mobility Through e-technologies

Pietro Belotti, Rossella Moscarelli, and Luca Pirolo

Abstract The response to the environmental crisis represents one of the main challenges of the mobility field. How should the mobility sector, related both to commuting transport and leisure travel, be reshaped in the contemporary era? Slow and electric mobilities can provide useful practices and technologies to tackle this complex issue. In particular, integrating slow mobility with electric vehicle supply equipment technologies (EVSE) appears as an interesting field of development. Starting from this perspective, the volume critically discusses the so-called slow e-mobility, which includes practices, systems, and technologies related mainly to pedal-assisted bicycles. The main topic of the work is then introduced by considering the recent cultural and academic debate. The book contributes to such reflections by offering a broad overview of the opportunities and risks of this innovative form of mobility, with specific attention to its spatial effects.

1 Sustainability and Mobility: An Urgent Issue

The mobility and transport field is living in a challenging era due to the necessity of responding to the climate crisis (Hickman and Banister 2007; Chapman 2007; Anderton 2010; Willson 2012), and relevant social changes (Castles 2018).

Environmental and ecological concerns are rising worldwide, as carbon emissions are increasing rapidly, which is one of the leading causes of global warming. These emissions come from different sectors, but transportation is the biggest contributor (Babar et al. 2020). From this perspective, low-emission transport is an increasingly

P. Belotti
Department of Electronics, Information, and Bioengineering, Politecnico di Milano, Milan, Italy
e-mail: Pietro.belotti@polimi.it

R. Moscarelli (✉)
Department of Architecture and Urban Studies, Politecnico di Milano, Milan, Italy
e-mail: Rossella.moscarelli@polimi.it

L. Pirolo
Department of Electronics, Information, and Bioengineering, Politecnico di Milano, Milan, Italy
e-mail: Luca.pirolo@polimi.it

© The Author(s), under exclusive license to Springer Nature Switzerland AG 2026
R. Moscarelli et al. (eds.), *The Future of Slow Mobility*, Research for Development,
https://doi.org/10.1007/978-3-032-23660-9_1

urgent issue that could be addressed by improving the technology of vehicles or by changing transport and travel habits.

As a consequence of the environmental crisis, demographic changes must also be considered as drivers of new trends in several fields, including mobility. In 2050, the Earth will have 9, 7 billion people (United Nations Organization 2022). Areas like Europe have an aging population: the over-65 will constitute about 27% of the population by 2050 (United Nations Organization 2022). Other areas in Africa and Asia are registering population growth (United Nations Organization 2022). Another challenge is urbanization: 7 persons out of 10 by 2050 will live in the cities and every day more than 172 thousand new inhabitants choose the cities to settle in (United Nations Organization 2019).

In sum, sustainable mobility is an urgent and necessary project theme to address the contemporary challenges of climate and ecological transition. First of all, because mobility is one of the sectors that contributes most to the pollution of our planet, through the emission of climate-altering gases, the consumption of natural resources, such as fossil fuels, in addition to air and noise pollution, and the congestion of the most populated urban areas. For some time now, people have been thinking about which forms of mobility could best contribute to making our journeys, both daily and leisurely, as environmentally friendly as possible. The "umbrella" concept of sustainable mobility, deriving from the general concept of sustainable development, encompasses the numerous practices that have been tested for decades to reconcile the right to mobility with the need to reduce the so-called negative externalities associated with it (Marioli 2013). *Smart*, *green*, *shared*, *electric*, *active,* and *soft* mobility are some of the increasingly numerous possible declinations of sustainable mobility. A carousel of terms and definitions, partly overlapping. Even if there is no shared and definitive way to define each practice, it is still useful to provide a synthetic glossary, as done in Table 1.

As shown in Fig. 1, it is possible to collect the numerous practices of sustainable mobility into two macrofamilies, defined starting from the speed of movement. Fast mobility includes the entire world of fast mobility, that is, moving at more than 20 km/h. Below this threshold, we instead speak of slow mobility. This speed threshold allows for the activation of a series of relationships between those who move and the context within which such movement occurs (Pileri and Moscarelli 2021).

Slow and fast mobility can therefore both contribute to the challenge of sustainable mobility, declining through the different green practices defined above. Both can do it, although it is necessary to note that some forms of fast mobility are not recognized as sustainable (for example the use of private fossil fuel cars).

This book focuses just on the slow mobility, regarded as a branch of sustainability (Pelgrims 2022; Munch 2024), which includes walking, cycling, running, and skating (Fullagar et al. 2012). It is also expressed as Non-Motorized Transportation (NMT) and characterized by means with a low environmental impact (Eldeeb and Tohlob 2022). More specifically, we aim to consider how two different practices of sustainable mobility can interact and influence each other, i.e. slow and e-mobility.

Table 1 Some definitions of different practices of sustainable mobility. *Source* Developed by the authors

Sustainable mobility practice	Author/s	Definition
Smart mobility	Staricco (2013: 342)	Mobility system characterised by a consistent and systematic use of technological innovations, both in terms of ICT (used to provide information to travellers, to streamline traffic, to manage public transport fleets, to improve freight transport logistics, etc.) and in terms of means of transport (electric cars, on-demand buses, bikes and car sharing, etc.)
	Docherty et al. (2018: 118)	It is characterized by the shift towards "mobility as a service", increasingly intelligent infrastructure, the electrification of the vehicle fleet and automated vehicles that do not require 'driving' by any of the passengers
Shared mobility	Shaheen et al. (2015: 4)	It includes various forms of carsharing, bikesharing, ridesharing (carpooling and vanpooling), and on-demand ride services. It can also include alternative transit services, such as paratransit, shuttles, and private transit services, called microtransit, which can supplement fixed route bus and rail services
	Shokouhyar et al. (2021: 1)	It is an emblem of a shared economy that is linked with both sustainable travel behavior and sustainable urban infrastructure from the social, economic, and environmental perspective
Green mobility	Almatar (2023: 3)	Mobility that does not depend on the factors that become a significant part of diminishing natural resources such as fossil fuels
	Echeverría et al. (2022: 1)	Green modes of mobility include a variety of eco-friendly alternatives, including active (walking or cycling) and public transport, as well as car-sharing, carpooling, and electric scooters, among others
E-mobility	Behrendt (2018: 3)	Short for electric mobility—is an umbrella term for mobilities involving electric vehicles, i.e. vehicles that rely on an electric motor for at least some of their propulsion with the power coming from an on-board battery (e.g. car or bike) or an off-vehicle source (e.g. train)
	Hoch et al. (2015: 455)	Vision of future transportation by means of a network of electric vehicles allowing people to fulfill their individual mobility needs in an environmental friendly manner (decreasing pollution, saving energy, sharing vehicles, etc.

(continued)

Table 1 (continued)

Sustainable mobility practice	Author/s	Definition
Active mobility	Maltese et al. (2021: 2)	It includes non-motorized transport activities, such as walking, cycling, and recent human-powered variants (e.g. push scooter, roller skates/rollerblades and skateboard)
	Haufe et al. (2016: 49)	It is not only healthy on the individual level, […], it is also carbon–neutral, cheap and more space saving than other modes of transport
Soft mobility	Fistola and Rocca (2018: 302)	Conscious approach to mobility, a type of mobility that includes all forms of non-motorized transport (NMT) that exclusively uses "human energy" (Human Powered Mobility)
	Galderisi and Ceudech (2008: 57)	Form of sustainable mobility that, while respecting the right of the individual to move, allows for the improvement of living conditions, influencing in particular: air and noise pollution; congestion generated by urban vehicular traffic; the need to improve road safety levels

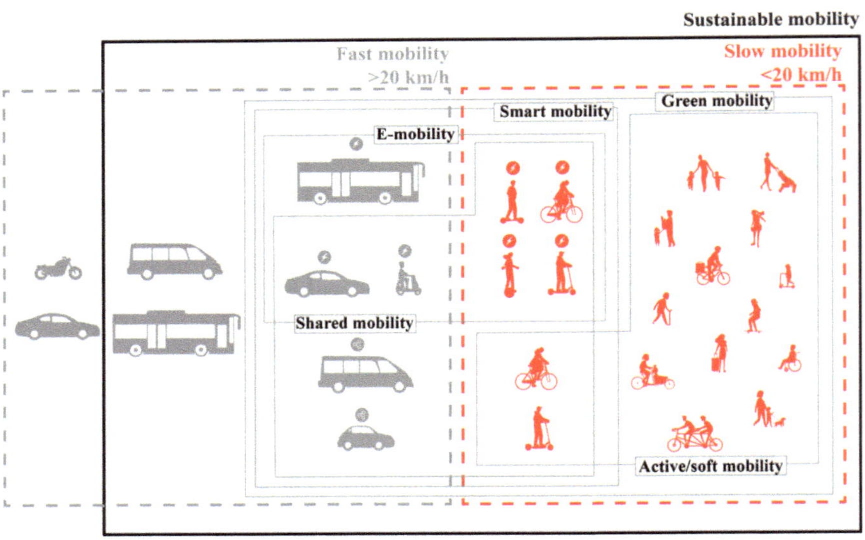

Fig. 1 Scheme of the different practices of sustainable mobility. Source: Developed by the authors

2 The Slow E-Mobility as an Essential Strategy of Sustainability

The so-called slow e-mobility, namely the integration of electric vehicle supply equipment (EVSE) technologies and slow mobility, represents a unique possibility in the reflection on the new era of sustainable mobility (Popovich et al. 2014; Fishman and Cherry 2015; Dill and Rose 2012; Rérat 2021) since it combines the positive effects of both slowness and electric technologies (Bourne et al. 2022; Berntsen et al. 2017).

Starting with the slow mobility, the advantages already described in the literature are many and quite wide: from pollution reduction (Zickgraf 2021) to the positive effects on health and lifestyle (Saunders et al. 2013); from the improvement of the quality of community life (Holden et al. 2020) to the enhancement of social inclusion (Harms et al. 2014). Finally, if slow mobility is enlarged to the dimension of leisure travel, it can also become a tourist resource, with many positive economic aspects and few environmental impacts (Pileri 2016).

Talking about electric mobility, the benefits to society are multiple and overcome the current transportation style of moving based on internal combustion engine (ICE) vehicles. There is an important improvement in the reduction of local pollutants since electric vehicles do not produce local pollutants, such as CO, NOx, and particulate matter, compared to ICE vehicles (European Environment Agency 2016). The electric motors are extremely efficient concerning ICE, which means less energy is spent for the same use. If the energy used to recharge electric vehicles comes from renewable sources, this can decrease the variation in energy price and energy security (Singh et al. 2023). For the users, electric vehicles have lower maintenance costs because they have fewer components, which can diminish the expenditures for the electric vehicles (Sanguesa et al. 2021). And finally, electric mobility diffusion is stimulating many sectors such as battery, recharging infrastructure, and renewable energy production. It is opening new job opportunities and economic development (Foley et al. 2010).

Overall, slow e-mobility has many relevant positive impacts, derived from the peculiarities of slowness and e-technologies. E-bikes also make it possible to extend access to slow mobility, widening the range of users, the variety of vehicles, and the places of use (Rodríguez Quintero and Vidal Abarca Garrido 2022; Corti et al. 2024).

On the other hand, it is necessary to consider also the main critical aspects. In part they are similar to those related to slow mobility, concerning: the speed itself, which is low so it takes a longer time to move quickly for daily commuting (Bassett et al. 2008), while extremely hot weather or heavy rains can affect the users' daily habits (Winters et al. 2010). Not all cities have good slow-mobility infrastructures, and this can be seen as an obstacle for potential users (Pucher et al. 2010), especially in terms of safety (Pucher and Buehler 2017). Some barriers can slow the adoption of these technologies (Corti et al. 2024). These weaknesses add to those related to the EVSE technology. Range anxiety is one of the best-known reasons to avoid using these technologies, followed by limited accessibility of all social classes, urban regulations, costs, public perception, and manufacturing and supply chains that are weak, and not equally spread worldwide (Offer et al. 2010). The recharge time is another factor that affects the growth of electric vehicle users. The recharge phase with state-of-the-art technology has a much longer duration compared to traditional vehicles (Neubauer and Wood 2014) and this type of infrastructure is not homogeneously distributed on a regional scale. Most of the time, the demand drives the positioning, leaving peripheral territories without infrastructure (Sierzchula et al. 2014). On the market, electric vehicles are more expensive than traditional ones, which can be a barrier to the diffusion of electric mobility (Nykvist and Nilsson 2015). Electric vehicles store energy inside batteries that are made with materials characterized as critical, i.e. non homogeneously present on the Earth. This can determine dependency on specific market players (Olivetti et al. 2017). The above issues create a cultural skepticism for the electric vehicle among users (Krishna 2021; Pamidimukkala et al. 2024).

3 Promoting Slow and E-Mobility

Many National and European policies are trying to reduce the mobility sector's environmental impacts so as to render Europe the first carbon-free continent by 2050, by giving particular emphasis to both slow and e-mobility.

The Sustainable and Smart Mobility Strategy, carried out by the European Commission in 2021, introduces e-mobility as the first pillar of mobility in the Green Deal era, where all transportation will be zero-emission by 2050 (European Commission 2021). Moreover, the European Commission adopted, in the second half of 2023, the 'Declaration of Cycling', categorizing all the various types of slow vehicles present in our society and fostering the usage of slow and electric transport systems such as e-bikes or pedelecs to create a more inclusive, sustainable, and right society. To do this, better and safer road and service infrastructure must be developed in the next years (European Commission 2023).

This approach is also assumed by the national policies, such as the Italian "Piano Nazionale Integrato per l'Energia e il Clima" (Ministero dello Sviluppo Economico, Ministero dell'Ambiente e della Tutela del Territorio e del Mare, Ministero delle Infrastrutture e dei Trasporti 2019), which contains the 'Piano Generale della Mobilità Ciclistica urbana e extraurbana', drafted in 2022 by the then Ministry of Sustainable Infrastructures and Mobilities (Ministero delle Infrastrutture e Mobilità Sostenibili 2022). This document concretely lays out the strategy to transform Italy into a country where the right to cycle is on par with other means of transport, giving policy guidelines on various areas such as urban and cycling tourism, and putting resources into large strategic projects and local-scale projects to complete an infrastructure that has been fragmented for too long. In the document, the cycle touristic infrastructure is defined, and it will patch up less-known and more isolated Italian territories, revitalizing hamlets, smaller historical centers, and less anthropized inland areas. Furthermore, in line with the European declaration, it promotes the use of all forms of cycling, considering electric vehicles as fundamental to the enlargement of cycling users and developing the cycling sector as an alternative and clean daily means of transport.

4 The Future of Slow Mobility: Aims, Originality, and Contents of the Book

The reflections of this volume arise from this policy context and from the actual scientific and public debate. The main aim is to consider how e-technologies can be integrated into the plan and design of a slow-mobility project, both about the daily habits in an urban context, as well as tourist movements. This implies questioning the impacts of slow e-mobility on the territory, namely on the places and on society.

How can the city and the territory change with the spread of e-bikes? What ethical implications are there in this innovative mode of mobility and travel? How can the public actor best plan this sustainable practice? How can sustainable and ecological mobility adapt to the design of the cycling infrastructure? These are the main research questions that the volume faces thanks to a wide overview of studies by researchers and experts with different disciplinary backgrounds: architects, urban planners, engineers, computer scientists, sociologists, economists, and philosophers. Such a perspective, focused on the complex and various territorial effects of slow e-mobility, is the main original point of the book. Other works (Han et al. 2020; Senese et al. 2023; Lagerstedt and Svensson 2022; Pröbstl-Haider et al. 2018; Gall et al. 2023; Pal and Kunj 2024; Toll 2021) tackle similar issues from different perspectives. Many observe electric mobility by focusing just on cars or public transportation (e.g. bus or tram); others consider the integration between mobility and urban planning and design without reflecting on the particular and challenging case of slow e-mobility. In another set of works, the topic of e-bikes is treated just as a technical improvement (subject of mechanical engineering) or in non-scientific books.

Planning the services of slow electric mobility is a topic that has not been exploited and studied in all aspects, to the best of our knowledge. This book can thus represent a first attempt to explore slow e-mobility as a means for both daily commute and leisure traveling, with a rigorous and academic approach, and by providing a synoptic view of several different aspects related to one another.

Our aim is to integrate technological aspects and service planning, understanding what is needed in urban and extra-urban contexts to spread this new way of transport.

Assuming such aims, the volume is articulated in two parts. The first part, "Stating the issue of slow and electric mobility", introduces the pillars of the topic and how slow e-mobility challenges urban and travel slow practices.

The first part begins with a reflection on the practice of mobility and tourism: mobility on e-bicycles is an inherently significant part of slow and light mobility, both when considered for daily commuting and within a touristic context. This first chapter covers a deep analysis by Fossa and Piazza of the ethics of using e-bikes as a promising means of transport for current and future mobility scenarios and the need to identify criticalities in how e-bike usage is regulated. In the second chapter by Colleoni and Jankowski examine the inclusivity perspective of slow electric mobility and the potential risks and opportunities of using e-bikes to create a more equitable paradigm of urban mobility. In Chap. 3, Carra and Pezzagno focus on the planning aspect of urban infrastructures, which needs solid guidelines to identify its risks and opportunities and to guarantee a clear transition toward greater use of sustainable means of transportation such as e-bicycles. In the fourth chapter, Pucci and Morais debate the actual positioning of charging infrastructure in the urban environment, with an emphasis on differentiating between e-bikes and electric vehicles (EVs), while highlighting the imbalance of charging capabilities by introducing the power gap, an objective indicator of the demand for public charging facilities. Pirolo and Pileri (Chap. 5) discuss the link between slow tourism and electric bikes, giving an interpretation of the phenomenon and the possible implications from an inclusivity and accessibility point of view. Fortini e Uccellatori end the first part by introducing the idea of "Landscape Metropolis", where the landscape is meant to be part of the infrastructure upon which a sustainable, capillary mobility network can be established, both to favor sustainable mobility and to improve the connectivity of more fragile territories.

The second part, "Planning and Designing E-Mobility", explores the technical dimensions of sustainable mobility, offering a broad perspective on both operational and strategic planning for urban and extra-urban electric slow mobility. Leoni and Tanelli (Chap. 7) investigate the development of e-bike functionalities capable of monitoring both the vehicle and the surrounding environment in order to enhance riding safety and comfort. The focus on safety continues in Chap. 8, where Raimondi et al. extend an established approach to assessing user safety during e-bike use and present a comprehensive methodology that includes operational measures aimed at improving cycling safety. Chap. 9, by Petaccia and Gungor, examines how mobility can shape public space in cities, reflecting on how urban environments might be transformed if mobility were predominantly slow and electric. Urban space is also increasingly influenced by last-mile delivery systems: Beatrici, Maggioni, Birolini, and Malighetti (Chap. 10) provide a review of sustainable and light urban delivery mobility, with a specific focus on the use of electric cargo bikes for fleet management and vehicle routing planning. In Chap. 11, Moscarelli explores the potential of integrating e-technology with slow tourism, analysing the installation process of Electric Vehicles Supply Equipment services in different European contexts, focusing on the location, reasons, aims, and implementation of the projects, as well as the private or public actors involved. The final two chapters delve into the mathematical foundations of planning methodologies for cycling networks. Pirolo, Moscarelli, and Belotti (Chap. 12) present an optimization model to support decision-makers—such as regional administrations—in increasing the attractiveness of cycleway networks through the strategic installation of e-bike charging stations, applying the model to the Sistema Nazionale delle Ciclovie Turistiche. Finally, Pirolo, Belotti and Malucelli (Chap. 13) extends an existing planning approach by employing optimization models that jointly address charging infrastructure deployment and the design of attractive cycling itineraries.

The main part of this publication concerns the multidisciplinary research developed within the MOST—Centro Nazionale per la Mobilità Sostenibile (National Centre for Sustainable Mobility), funded by the European Union Next-GenerationEU. MOST, through collaboration with 24 universities, the National Research Council (CNR) and 24 large companies, has the mission to encourage and support the development of modern, sustainable and inclusive mobility solutions for the entire country, ranging from air mobility, light vehicles, and active mobility, waterborne transport, rail, as well as the research on new fuels such as hydrogen, digital systems for accident reduction, among others, to make the mobility system 'greener' as a whole and more 'digital' in its management (Centro Nazionale per la Mobilità Sostenibile, 2023).

Author Contributions and Acknowledgments The authors have shared the concept of the article. P.B. wrote Sect. 4. L.P. wrote Sects. 2–3. R.M. wrote Sects. 1.

The paper is a result of the study carried out within the MOST—Sustainable Mobility National Research Center and received funding from the European Union Next-Generation EU (PIANO NAZIONALE DI RIPRESA E RESILIENZA (PNRR)—MISSIONE 4 COMPONENTE 2, INVESTIMENTO 1.4—D.D. 1033 17/06/2022, CN00000023). This chapter reflects only the authors' views and opinions; neither the European Union nor the European Commission can be considered responsible for them.

References

Almatar KM (2023) Towards sustainable green mobility in the future of Saudi Arabia cities: implication for reducing carbon emissions and increasing renewable energy capacity. Heliyon 9(3):3

Anderton K (2010) Two billion cars: driving towards sustainability. Transp Rev 30:675–676

Babar AHK, Ali Y, Khan AU (2020) Moving toward green mobility: overview and analysis of electric vehicle selection, Pakistan a case in point. Environ Dev Sustain 23:10994–11011

Bassett DR, Pucher J, Buehler R, Thompson DL, Crouter SE (2008) Walking, cycling, and obesity rates in Europe, North America, and Australia. J Phys Act Health 5:795–814

Behrendt F (2018) Why cycling matters for electric mobility: towards diverse, active and sustainable e-mobilities. Mobilities 13(1):64–80

Berntsen S, Malnes L, Langåker A, Bere E, (2017) Physical activity when riding an electric assisted bicycle. Int J Behav Nutr Phys Act 14

Bourne, JE, Kelly P, Mutrie N (2022) The rise of the electrically assisted bicycle and the individual, social and environmental impacts of use. Cycling 27–64

Castles S (2018) Social transformation and human mobility: reflections on the past, present and future of migration. J Intercult Stud 39:238–251

Centro Nazionale per la Mobilità Sostenibile (2023) MOST (Centro Nazionale per la Mobilità Sostenibile). https://www.centronazionalemost.it/. Accessed 29 Oct 2024

Chapman L (2007) Transport and climate change: a review. J Transp Geogr 15:354–367

Corti F, Iacono SD, Astolfi D, Pasetti M, Vasile A, Reatti A, Flammini A (2024) A comprehensive review of charging infrastructure for electric micromobility vehicles: technologies and challenges. Energy Rep 12:545–567

Dill J, Rose G (2012) Electric bikes and transportation policy: insights from early adopters. Transp Res Record J Transp Res Board 2314:1–6

Docherty I, Marsden G, Anable J (2018) The governance of smart mobility. Transp Res Part A Policy Pract 115(114–125):118
Echeverrìa L, Gimenez-Nadal JI, Molina JA (2022) Green mobility and well-being. Ecol Econ 195:1
Eldeeb A, Tohlob A (2022) Waterfront Slow-motility development (Mansoura—Egypt). ERJ. Eng Res J
European Commission (2021) The sustainable and smart mobility strategy
European Commission (2023) European declaration of cycling
European Environment Agency (2016) Electric vehicles and the energy sector: impacts on Europe's future emissions. Publications Office
Fishman E, Cherry C (2015) E-bikes in the mainstream: reviewing a decade of research. Transp Rev 36:72–91
Fistola R, La Rocca RA (2018) Slow mobility and cultural tourism. Walking on historical paths. In: Smart planning: sustainability and mobility in the age of change, pp 301–322
Foley AM, Winning IJ, O Gallachoir BP (2010) State-of-the-art in electric vehicle charging infrastructure. IEEE
Fullagar S, Wilson E, Markwell K (2012) Starting slow: thinking through slow mobilities and experiences. Slow Tourism Multilingual Matters 1–9
Galderisi A, Ceudech A (2008) Mobilità dolce e reti pedonali in città. TeMA-J Land Use Mobility Environ 1(3):57
Gall T, Vallet F, Reyes Madrigal LM, Hörl S, Abdin A, Chouaki T, Puchinger J (2023) Sustainable urban mobility futures. Springer Nature Switzerland
Han H, Lho LH, Al-Ansi A, Yu J (2020) Cycling tourism: a perspective article. Tourism Rev 75:162–164
Harms L, Bertolini L, te Brömmelstroet M (2014) Spatial and social variations in cycling patterns in a mature cycling country exploring differences and trends. J Transp Health 1:232–242
Haufe N, Millonig A, Markivica K (2016) Developing Encouragement Strategies For Active Mobility. Transp Res Procedia 19(49–57):49
Hickman R, Banister D (2007) Looking over the horizon: transport and reduced CO_2 emissions in the UK by 2030. Transp Policy 14:377–387
Hoch N et al (2015) The E-mobility case study. In: Software engineering for collective autonomic systems: the ASCENS approach, 513–533
Holden E, Banister D, Gössling S, Gilpin G, Linnerud K (2020) Grand narratives for sustainable mobility: a conceptual review. Energy Res Soc Sci 65. https://doi.org/10.1016/j.erss.2020.101454
Krishna G (2021) Understanding and identifying barriers to electric vehicle adoption through thematic analysis. Transp Res Interdisc Perspect 10
Lagerstedt E, Svensson H (2022) Do tourists dream of electric bikes? In: Electric bikes as a mean to improve sustainability of tourism in rural Sweden, pp 167–178
Maltese I, Gatta V, Marcucci E (2021) Active travel in sustainable urban mobility plans an Italian overview. Res Transp Bus Manag 40:2
Marioli L (2013) Mobilità sostenibile e trasporto intermodale. Rivista di Diritto Dell'economia, Dei Trasporti e Dell'ambiente 11:19–39
Ministero delle Infrastrutture e Mobilità Sostenibili (2022) Piano Generale della Mobilità Ciclistica
Ministero dello Sviluppo Economico, Ministero dell'Ambiente e della Tutela del Territorio e del Mare, Ministero delle Infrastrutture e dei Trasporti (2019).Piano Nazionale Integrato Energia e Clima
Munch E (2024) Sustainable and slow cities: the quest for conviviality. Sustain Sci Pract Policy 20
Neubauer J, Wood E (2014) The impact of range anxiety and home, workplace, and public charging infrastructure on simulated battery electric vehicle lifetime utility. J Power Sources 257:12–20
Nykvist B, Nilsson M (2015) Rapidly falling costs of battery packs for electric vehicles. Nat Clim Chang 5:329–332

Offer GJ, Howey D, Contestabile M, Clague R, Brandon NP (2010) Comparative analysis of battery electric, hydrogen fuel cell and hybrid vehicles in a future sustainable road transport system. Energy Policy 38:24–29
Olivetti EA, Ceder G, Gaustad GG, Fu X (2017) Lithium-ion battery supply chain considerations: analysis of potential bottlenecks in critical metals. Joule 1:229–243
Pal K, Kunj T (2024) Advancements in Electric vehicle infrastructure: from development to optimization: a comprehensive guide to optimizing EV infrastructure. River Publishers
Pamidimukkala Kermanshachi S, Rosenberger JM, Hladik G (2024) Barriers and motivators to the adoption of electric vehicles: a global review. Green Energy Intell Transp 3
Pelgrims C (2022) Towards ethical mobility behavior: the socio-ecological potentiality of ambiance-based mobility infrastructure design. In: Structures and architecture a viable urban perspective? CRC Press, pp 527–532
Pileri P (2016) Vento the gentle revolution cycling its way through the landscape. Corraini Edizioni
Pileri P, Moscarelli R (2021) Slow Tourism to slow travel: an idea for marginal regions. In: Pileri P, Moscarelli R (eds) Cycling & walking for regional development. How slowness regenerates marginal areas. Springer Nature Switzerland, pp 23–38
Popovich N, Gordon E, Shao Z, Xing Y, Wang Y, Handy S (2014) Experiences of electric bicycle users in the Sacramento, California area. Travel Behav Soc 1:37–44
Pröbstl-Haider U, Lund-Durlacher D, Antonschmidt H, Hödl C (2018) Mountain bike tourism in Austria and the Alpine region–towards a sustainable model for multi-stakeholder product development. J Sustain Tour 26:567–582
Pucher J, Buehler R (2017) Cycling towards a more sustainable transport future. Transp Rev 37:689–694
Pucher J, Dill J, Handy S (2010) Infrastructure, programs, and policies to increase bicycling: an international review. Prev Med 50:106–125
Rérat P (2021) The rise of the e-bike: towards an extension of the practice of cycling? Mobilities 16:423–439
Rodríguez Quintero R, Vidal Abarca Garrido C (2022) Revision of the EU green public procurement criteria for road transport
Sanguesa JA, Torres-Sanz V, Garrido P, Martinez FJ, Marquez-Barja JM (2021) A review on electric vehicles: technologies and challenges. Smart Cities 4:372–404
Saunders LE, Green JM, Petticrew MP, Steinbach R, Roberts H (2013) What are the health benefits of active travel? A systematic review of trials and cohort studies. PLoS ONE 8. https://doi.org/10.1371/journal.pone.0069912
Senese A, Pelfini M, Maragno D, Bollati IM, Fugazza D, Vaghi L, Federici M, Grimaldi L, Belotti P, Lauri P, Ferliga C, La Rocca L, Diolaiuti GA (2023) The role of e-bike in discovering geodiversity and geoheritage. Sustainability 15(6). https://doi.org/10.3390/su15064979
Shaheen S, Nelson C, Apaar B, Cohen A (2015) Shared mobility: a sustainability & technologies workshop: definitions, industry developments, and early understanding, p 4
Shokouhyar S, Shokoohyar S, Sobhani A, Jafari Gorizi A (2021) Shared mobility in post-COVID era: new challenges and opportunities. Sustain Cities Soc 67:1
Sierzchula W, Bakker S, Maat K, van Wee B (2014) The influence of financial incentives and other socio-economic factors on electric vehicle adoption. Energy Policy 68:183–194
Singh D, Paul UK, Pandey N (2023) Does electric vehicle adoption (EVA) contribute to clean energy? Bibliometric insights and future research agenda. Cleaner Responsible Consumption 8
Staricco L (2013) Smart mobility: opportunità e condizioni. TeMA J Land Use Mobility Environ 6(3):342–354
Toll M (2021) The electric bike manifesto: how electric bicycles can save our world (if we let them). Toll Publishing
United Nations Organization (2019) World urbanization prospects the 2018 revision. United Nations
United Nations Organization (2022) World population prospects 2022 summary of results. United Nations

Willson R (2012) Transport revolutions: moving people and freight without oil. J Urban Technol 19:106–107
Winters M, Davidson G, Kao D, Teschke K (2010) Motivators and deterrents of bicycling: comparing influences on decisions to ride. Transportation 38:153–168
Zickgraf C (2021) Climate change, slow onset events and human mobility: reviewing the evidence. Curr Opin Environ Sustain 50:21–30

Stating the Issue of Slow and Electric Mobility

Ethics of (e-)Cycling: A Preliminary Overview

Fabio Fossa and Alessandro Piazza

Abstract This chapter aims to discuss the uptake of (e-)bikes from the viewpoint of the ethics of transportation. More specifically, it intends to identify and critically examine the moral values and arguments that are commonly mobilized when (e-) cycling is presented as a promising means of transportation for current and future mobility scenarios. Two lines of inquiry are explored. First, e-bikes, being a hybrid of active and motorized means of transportation, interestingly intersect calls for both modal shift towards cycling and transition to e-mobility solutions. The two visions, however, do not necessarily rest on shared ethical grounds, which might generate confusion and, thus, require further clarification. Secondly, projects aimed at fostering the use of (e-)bikes in urban and non-urban contexts also rest on value frameworks that only partially overlap, which equally poses the need for a critical examination of their different normative presuppositions. By studying the ethics of (e-)cycling along these two axes, the contribution endeavours to provide a critical discussion of the moral desiderata linked to this transport mode and, thus, to support ethical awareness in its integration into current and future transportation systems.

1 Introduction: Ethical Values and the Role of Cycling in Contemporary Transport

Values are woven into nearly every aspect of transport debates. Whether we are advocating for road safety, sustainable mobility, or equitable access, we are making judgments about which values should orient the development of our transport systems.

F. Fossa (✉) · A. Piazza
Department of Mechanical Engineering, Politecnico Di Milano, Milan, Italy
e-mail: fabio.fossa@polimi.it

A. Piazza
e-mail: a.p.piazza1@tue.nl

A. Piazza
MOVEMENT Research Group, Department of Mechanical Engineering, Eindhoven University of Technology, Eindhoven, Netherlands

© The Author(s), under exclusive license to Springer Nature Switzerland AG 2026
R. Moscarelli et al. (eds.), *The Future of Slow Mobility*, Research for Development,
https://doi.org/10.1007/978-3-032-23660-9_2

Some values have historically been more widely recognized. Safety, for example, has long concerned transport experts, while sustainability has more recently become central to transport practice, with its composite call to social, economic, and environmental well-being. Other values, such as privacy, autonomy, and justice, are increasingly attracting the attention of scholars, practitioners, and activists alike.

But what is a value? Values articulate shared conceptions of what is good, desirable, or worth striving for. Moreover, values are not just abstract ideals, but play a crucial role in guiding our action: they direct policy choices, shape priorities and investments, and influence how social systems develop. In this sense, values embed a vision of the future, of the kind of society we aim to create, and the kind of lives we think people should be able to live. Ethical values, therefore, fundamentally inform transport decision-making, shaping the kind of mobility systems we strive to build. Nonetheless, a structured reflection on the ethics of transportation is yet to be fully established as a field of study (Richardson 1995; Santoni de Sio 2021). This chapter aims at contributing to this process by focusing on how ethical values have influenced the debate on (e-)cycling, addressing for the first time this issue in a systematic and reflective way.

Indeed, cycling has long been inspiring perspectives on transportation that are profoundly imbued with ethical values. Classic works such as Ivan Illich's *Energy and Equity* (Illich 1974) and popular books such as David Byrne's *Bicycle Diaries* (Byrne 2021) articulate visions where cycling is decisively characterized as capable of promoting individual and social well-being in a context so fraught of conflicts such as urban transport and life. The intimate connection between cycling and ethical values is also noticeable in current frameworks that integrate urban and transportation planning to transition from car-centric cities to more liveable, human-centric environments.

The rise of e-bikes is not only transforming our transportation systems—it is also reshaping the ethical discourse around cycling. As this chapter shows, cycling has traditionally been linked to values like sustainability, health, and fairness in urban and non-urban life. E-bikes, by resembling conventional bikes but adding motorized assistance, concurrently reinforce and disrupt these ethical associations. This calls for a reassessment of how we understand the moral significance of cycling in today's transportation landscape.

As anticipated, this chapter presents some preliminary considerations on the uptake of e-cycling from the viewpoint of transportation ethics. More specifically, it aims to identify and critically discuss the moral values and arguments that are commonly mobilized when cycling is presented as a promising means of transportation for current and future mobility scenarios. Moreover, it intends to assess whether the uptake of e-bikes would confirm or challenge these ethical beliefs.

Two lines of inquiry are explored. First, as a hybrid of active and motorized transport, e-bikes sit at the intersection of calls to promote a modal shift toward cycling and to advance e-mobility solutions. However, these two mobility agendas do not necessarily share the same ethical foundations, which can lead to confusion and highlights the need for further clarification. Second, initiatives promoting e-bike use in urban and non-urban areas are based on different, sometimes only partially overlapping value systems. These differences indicate the importance of explicitly articulating and critically examining the values guiding these initiatives. By looking at the ethics of e-bikes along these two axes, we endeavour to provide a preliminary discussion of the moral desiderata linked to this means of transportation and, thus, to support ethical awareness in its integration into current and future transportation systems.

The chapter is structured as follows. Section 2 briefly overviews the wider urban and transportation frameworks in which ethical claims surrounding cycling are most notably embedded. The subsequent sections zoom in on relevant ethical values and consider how e-bikes might impact these expectations. The values we discuss are safety (Sect. 3), health (Sect. 4), environmental sustainability (Sect. 5), and justice (Sect. 6). A conclusion summarizes the results of the study.

2 Cycling and the Car-Lite City

As mentioned, current claims linking cycling to ethical values are commonly embedded in urban visions aimed at contrasting car-centric design and promoting cities where human well-being is given priority. Frameworks sharing this objective come under various names, but all look at bicycles as catalysts of their agenda.

In this sense, the most well-known framework is perhaps Carlos Moreno's *15-minute City*, which advocates for restructuring urban environments along the lines of proximity, density, mixed use, and ubiquity. The approach strives to redesign cities so that services essential to well-being would be reachable in 15 min by walking, biking, or public transport, thus "reduc(ing) road congestion, pollution, and dependence on the private vehicle, while promoting an active and healthy lifestyle for residents" (Moreno 2015, p. 105).

In his book *The Bicycle City*, Dan Piatkowski (2024) echoes these positions by advocating for the *car-lite* city—i.e., a city where car traffic is not necessarily banned, but cleaner, healthier, and more socially cohesive options are given priority. Bicycles, Piatkowski stresses, are just one ingredient of the transition, but an eminent one: a catalyst that, if opportunely and intelligently added to the mix, can bring substantial results at limited costs.

Similarly, the *Bicycle Urbanism* movement—so well described in Colville-Andersen's book *Copenhagenize* (Colville-Andersen 2018)—also stresses the potential of cycling in the struggle against car dependency and for more liveable, human-centric urban environments. Even if bicycles are obviously only a part of the bigger picture, these perspectives on the future of cities share a vision according to which promoting cycling will bring us closer to urban environments that support, instead of hindering, human well-being.

As it is evident, these visions are based on normative foundations that define what an ethically desirable city should look like. Despite their differences, they often converge on key assumptions about well-being and moral values. Aspects that hinder urban dwellers' well-being are commonly identified in road safety risks, health issues, air and noise pollution, CO_2 emissions, extreme weather events linked to climate change, social fragmentation and isolation, lack of access to essential facilities and services, urban degradation, forced travel, and the social harms and economic costs associated with congestion. Private cars, and the cities shaped to accommodate them, are central to all these issues.

By reducing reliance on cars and prioritizing walking, cycling, and public transit—so these authors claim—urban space and transport systems can be rethought to better cope with many of the issues mentioned. Indeed, the path towards a safer, healthier, happier, cleaner, greener, more just and inclusive, more socially cohesive and resilient—in a word, more sustainable—urban life is importantly impacted by how people are enabled and encouraged to move (Henderson 2013; Montgomery 2013; Golub et al. 2016). In this journey, the role of cycling is acknowledged as essential.

E-bikes make their entrance against this background. It is against this background, then, that their ethical significance is to be assessed. The following sections take a closer look at the main ethical values informing narratives weaving together opposition to car-centrism, cycling, and liveable cities: safety, health, environmental sustainability, and justice. We do so with a twofold aim. First, we intend to critically discuss the association of these ethical values to cycling, so to shed a clearer light on what is at stake. Secondly, we assess how the advent of e-bikes challenges or reconfigures these associations, raising new questions about the ethical meaning of (e-)cycling in contemporary mobility systems.

We hope that this preliminary inquiry on cycling and transportation ethics might contribute to clarifying the moral expectations often linked to enabling and encouraging the use of bicycles both in and beyond cities.

3 Safety: Between Promise and Peril in a Car-Centric World

When cycling is discussed, the ethical value of safety is often mentioned in ways that starkly (and instructively) reveal the challenges and conflicts involved in efforts to overcome car dependence.

On the one hand, bicycles are promoted as means of transport that contribute to make streets safer (Jacobsen and Rutter 2012; Colville-Andersen 2018). Being light and relatively slow, they allow for gentle journeys which pose comparatively low risks. Furthermore, they are associated with traffic-calming effects, slowing down motor vehicles and creating a safer road environment for all—especially for vulnerable road users. Finally, as Piatkowski (2024, p. 3) notes, “by encouraging biking, neighborhoods can become denser, more walkable, more transit friendly”, which also contributes to minimizing the safety risks posed by the major cause of harm—drivers. On the other hand, cycling in cities is also widely associated with danger and risk. Not only statistics on traffic fatalities show that cyclists are exposed to a disproportionate risk of being harmed (de Hartog et al. 2010), but biking is often perceived as "terrifying" (Piatkowsky 2024, p. 28) and hazardous. How could cycling be conducive to safer streets if those who pedal are disproportionately exposed to harm and tend to feel unsafe?

Actually, the contrast between bicycles promoting safer streets and cyclists facing high risks is ill-conceived since it overlooks the crucial role of infrastructure (Elvik 2021). Current approaches to road safety such as Vision Zero (Björnberg et al. 2022) and the Safe System Approach (Marshall 2024) have demonstrated the importance of infrastructure design to enable road safety and influence transport mode preferences. Indeed, the contradiction looming on safety and cycling is a symptom of a lack of appropriate infrastructure for accommodating cyclists' needs (and rights). Their dire safety conditions are the result of a "transportation system we've given them but not designed for them" (Marshall 2024, p. 270). Car-centric urban and transportation planning have shaped an environment that is tailored around drivers' needs at the expenses of the safety and needs of other road users—especially cyclists and pedestrians.

For this reason, marking cycling as unsafe without highlighting the role of car-centric infrastructure is unconvincing at best. Encouraging the uptake of cycling in cities that are designed to prioritize driving without implementing the right cycling infrastructure, instead, is irresponsible. Indeed, building an environment that is safe and welcoming for cyclists is a prerequisite to properly measuring the traffic safety benefits that cycling yields. There is no contradiction, then, between *current* low levels of safety and high safety *potential*. Current low levels of safety only prove that safety is as much (if not even more) a matter of infrastructure than a matter of individual behaviour and preferences. Material measures such as protected bike lanes and policies such as those adopted within the framework of the *30 km/h City* constitute the conditions for safe cycling, without which any evaluation of its safety is partial at best.

This becomes even more important when e-bikes enter the picture. Although limited and somewhat mixed (Sundfør et al. 2020), evidence seems to suggest that e-bikes, being both heavier and faster, are harder to control and therefore more dangerous than regular bicycles (Schepers et al. 2014; Gross et al. 2018; Gehlert et al. 2018; Berk et al. 2022; Kroes et al. 2025). Such a relative increase in risk makes it even more evident that infrastructure is needed to separate traffic and protect different categories of vulnerable road users not just from drivers, but also from each other. And since road space is often unevenly distributed in favour of drivers, creating the conditions for safe cycling—and, thus, safer streets—means *redistributing* space from cars to bicycles (and, perhaps, other micromobility solutions). Conflict, then, is unavoidable. It represents an aspect of the discourse on safety and (e-)cycling that should neither be downplayed nor underestimated but expected and proactively managed.

A final consideration could be dedicated to tourism. e-Bikes are often praised for making cycling tourism more inclusive. As Santilli and Soldavini (2019, p. 72) envision, for example, mountain e-bikes will make mountain trails that are now enjoyable only by fit, expert hikers, more accessible to a much wider public. The consequences of such a widening of accessibility to territories where safety risks are tangible, that require expertise and care to be properly handled, and where rescue and first aid services are not always easy to deploy, must be assessed. Increases in accessibility are not to be paired with increased safety risks.

4 Health: Towards Inclusive Access to Active Mobility

The health benefits of cycling are central to its ethical appeal as a mode of transport (de Hartog et al. 2010; Garrard et al. 2012; Mueller et al. 2015; Santilli and Soldavini 2019; Garrard et al. 2021). Increasing sedentary lifestyles pose health threats that car-centric transportation models reinforce and are unfit to address. Active modes of transport can considerably help contrast sedentarism and deliver at least part of the exercise people need to stay healthy. The benefits of cycling on both physical and psychological health are many. Cycling regularly is associated with beneficial effects on the cardiovascular system, the cardiorespiratory system, muscle tone, stamina, resistance to fatigue, lower levels of stress, self-perceived well-being, and happiness due to environmental stimulation and endorphin release. Moreover, pedalling is a movement that enables aerobic exercise without burdening joints excessively.

Urban environments, however, could be unwelcoming to people who would like to travel and exercise at the same time. Even though it seems that pollution and poor air quality do not actually pose a higher risk of respiratory diseases to cyclists compared to other road users (e.g., Garrard et al. 2012; Tainio et al. 2016), they make urban cycling less pleasant and harder to perceive as a healthy practice. Here as well, the importance of implementing the right infrastructure to protect bicycle traffic and, at the same time, create the conditions for less polluted streets becomes visible.

The arrival of e-bikes has sometimes been seen as a step back in this sense. Being motorized and, therefore, limiting the 'active' component of cycling, doubts have circulated on the impacts of their uptake on health. The bias that e-cycling is 'cheating' has taken root.[1] Indeed, e-bikes that do not require input from users to move can be equated to passive modes of transportation. By offering an alternative that does not include exercise, they might jeopardize the health benefits of cycling.

e-Bikes endowed with pedal-assist, however, have been shown to have positive effects on health (Castro et al. 2019; Santilli and Soldavini 2019; Piatkowsky 2024). Even though the electric motor helps exercise force on pedals, users still have to contribute, which not only allows for a workout that is still significant in terms of health benefits, but also enables cycling for social groups who perceive regular cycling as out of their reach. Elderly people who lack the strength to pedal, unfit or overweight people who are looking for ways to integrate some forms of gentle

[1] See, e.g., https://corratec-uk.co.uk/do-you-get-any-exercise-with-an-e-bike-or-is-it-cheating and https://www.cyclinguk.org/article/why-riding-e-bike-or-e-cycle-good-you#:~:text=Additionally%2C%20e%2Dcycling%20can%20improve,potential%20alternative%20to%20conventional%20cycling".

exercise in their daily life, and other social groups who would greatly benefit from physical activity are now offered a new mode of transport that might help get the workout they need.

Moreover, e-bikes present the possibility to keep on exercising through cycling in less favourable conditions—e.g., in hilly territories or with hot weather, which in many cases is becoming more frequent due to climate change and might not only discourage regular cyclists, but also make it more dangerous for their health. As Piatkowsky (2024, p. 42) summarizes, "topography, temperature, ability, and fitness become less significant factors on an e-bike". Far from turning an active mode of transport into a passive one, e-bikes enable cycling to extend beyond the group of conventional cyclists, thereby bringing related health benefits to social groups that have traditionally been excluded.

5 Environmental Sustainability: Between Promise and Trade-Offs

In addition to safety and health, the value of environmental sustainability constitutes an essential piece of perspectives that insist on the ethical desirability of cycling. Indeed, the bicycle is the sustainable means of transport par excellence. Its manufacturing is relatively simple, does not require rare or contested materials, and produces low emissions. When used, it does not pollute, it is quiet and occupies very little space compared to alternatives. It is a long-lasting artefact that can be repaired and recommissioned, and does not become obsolete too easily. When it comes to air quality, CO_2 emissions, noise pollution, land use, life-cycle emissions, and durability, the environmental advantages of the bicycle as an alternative to the private car are clear and considerable.

Innovations in bicycle design have both amplified and challenged this narrative (Piatkowski 2024). For example, cargo bikes make cycling even more feasible as a replacement to both private and commercial motor vehicles. By allowing riders to safely transport extra weight, cargo bikes represent a new solution for transporting both other persons—e.g., kids—and goods—e.g., groceries. Moreover, cargo bikes offer a viable alternative to vans for delivering goods in cities—an opportunity that delivery companies are exploring in cities worldwide. The traffic-reducing, space-freeing potential of transforming many private and commercial trips from drives to rides thanks to cargo bikes is widely acknowledged.

Encouraging cycling in the pursuit of environmental goals not only applies to urban mobility. Municipalities around the world (PIARC 2023) have been working on networks of bike lanes to connect the city centre with the peripheries and hinterlands in an effort to decongest highways and offer more sustainable alternatives for commuting back and forth. Moreover, long-distance bike lanes crossing through territories also have potential to support more sustainable and active tourism and entertainment.

The advent of e-bikes supports these developments in many ways. The attractiveness of long-distance rides for both fun and transport is expected to increase with faster and less physically demanding vehicles. Cargo e-bikes make it easier and less strenuous to transport people or goods around city streets, simplifying not only the organization and execution of good delivery, but also the adoption of cargo e-bikes as replacements for private cars (Piatkowski 2024).

On the other hand, e-bikes come with higher environmental costs. Compared to regular bicycles, their production is more complex and more polluting. Materials for batteries are rare, contested, and extracted with significant environmental costs. e-Bikes' lifespan is shorter, their maintenance more complicated, their disposal more burdensome. The picture changes, however, if they are compared to cars and other motor vehicles (Huang et al. 2022; Philips et al. 2022). Apparently, then, the environmental desirability of e-bikes rests—perhaps not entirely, but importantly—on their ability to reduce the use of motor vehicles. This is important to consider, we believe, since some evidence suggests that e-cycling might sometimes replace cycling or public transport use rather than driving (Kroesen 2017).

Finally, an evaluation of environmental impacts should not be blind to the effects of infrastructure, which, as we have seen, constitutes an essential enabler of (e-) cycling. While this might be less relevant in urban areas, it definitely is in the case of natural environments. Considerations surrounding both the impacts of infrastructure (e.g., charging stations) and the effects linked to increased numbers of human visitors should be carefully evaluated when assessing the environmental sustainability of a project. Moreover, it is unclear whether similar projects should also be evaluated with regard to their potential to reduce tourism by private cars, or if in this case the environmental sustainability of a project should be decoupled from this objective.

6 Justice: Who Gets to Ride? Equity and Inclusion in (e-) Cycling

When it comes to the value of justice, a rapidly growing literature highlights how cycling infrastructure reflects and often reinforces broader patterns of social and spatial inequality (Doran et al. 2021; Martens et al. 2021; Tortosa et al. 2022; Cunha and Silva 2023; Cunha et al. 2024). Investments in cycling infrastructure often favour already privileged areas, while lower-income neighbourhoods remain underserved, limiting access to services and opportunities and to the broader benefits of active mobility for marginalized communities (Cunha et al. 2024). This pattern compounds existing social injustices, as disadvantaged groups tend to suffer from transport poverty, with limited car ownership and mobility budgets, and are often forced to rely on cycling as a cheaper alternative (Allen and Farber 2019; Doran et al. 2021). Moreover, the lack of safe and inclusive cycling infrastructure often means that disadvantaged groups are either excluded from the health benefits of active travel or exposed to greater risks in traffic when cycling in unsafe conditions

(Doran et al. 2021). Addressing these disparities requires not only more equitable infrastructure investments but also a broader cultural shift: embracing cycling as a mainstream and accessible mode of transport for all.

Indeed, in European countries with an established culture of cycling, disparities in cycling use tend to be smaller, supported by safer, more inclusive infrastructure that encourages participation across all groups, regardless of gender, income, education, or occupation (Tortosa et al. 2022). By contrast, in many developed but "low-cycling" countries—such as UK, Canada, and Australia—bicycle use is more prevalent among higher-income men and tends to correlate with higher education and stable employment, influenced by factors such as bicycle availability, secure storage, and the absence of stigma attached to cycling (Heesch et al. 2015; Fuller and Winters 2017; Tortosa et al. 2021). Moreover, cycling often correlates with recreational use for higher-income groups in these countries, while lower-income segments of the population tend to engage more in utility cycling, typically out of necessity. This trend becomes more pronounced in the US and South America, with lower-income groups making up a majority of cyclists – not as a result of inclusive planning, but due to more severe levels of transport poverty and social inequality (Tortosa et al. 2022). Despite these regional variations, disadvantaged communities everywhere tend to face structural barriers to cycling, including limited access to secure bicycles, poor-quality infrastructure, overexposure to pollution and traffic risks, and a lack of bike-sharing schemes. Improving access to cycling is, therefore, a crucial step toward advancing just transport systems and addressing the structural disadvantages faced by marginalized groups.

In this context, e-cycling holds the potential to either reduce existing disparities by improving accessibility and active mobility for disadvantaged groups or, if not carefully managed, to widen inequalities by mostly benefiting already privileged groups (Ballo et al. 2023). In fact, e-bikes differ from conventional bicycles in several ways. They allow higher speeds (up to 40 km/h), require significantly less physical effort, and enable users to travel longer distances. Their extended range supports uses beyond short trips for leisure or fitness, encouraging commuting, local deliveries, and even replacing some car trips. However, these advantages come with trade-offs, including higher costs and greater safety risks. These differences are relevant from a justice perspective, as they shape who can benefit from e-bikes and under what conditions.

Indeed, e-cycling raises important concerns about the distribution of the benefits and burdens of mobility. e-Bikes have the potential to reduce physical barriers to cycling, making active transport more accessible for older adults, the youth, people with health conditions, or those facing long or hilly commutes. They provide a more affordable alternative to car ownership for lower-income households by reducing costs, especially when supported by vouchers, subsidies, or tax credits. Additionally, e-bikes enable faster and farther trips, improving access to jobs and services in areas with limited public transport. However, e-bikes also risk widening inequalities when not carefully managed. Their higher cost means that without targeted financial support, they remain largely accessible only to wealthier users, widening existing socioeconomic inequalities in mobility. Moreover, the lack of safe cycling infrastructure in marginalized and low-income neighbourhoods increases safety risks, especially given the higher speeds and weight of e-bikes. Finally, many e-bike sharing programs tend to concentrate in central, high-demand areas while underserving peripheral neighbourhoods, reinforcing spatial injustices and denying equitable access to many disadvantaged communities (McNeil et al. 2017; Brown et al. 2019; Giuffrida et al. 2023; Berke et al. 2024).

Next to these distributive concerns, it is also essential to consider questions of procedural and epistemic justice in promoting e-cycling through policy and planning. Achieving procedural justice means involving all community members in the decision-making process. In this respect, participatory processes, such as citizen action groups and participatory budgeting, can go a long way in democratizing planning and supporting the creation of (e-)cyclable cities for all (Doran et al. 2021). At the same time, epistemic justice calls for greater awareness of the limitations of dominant planning tools, such as peak-hour bicycle counters or smartphone-based data, which often fail to capture the cycling patterns and needs of underrepresented groups. Planners should start instead from the lived experiences of people to assess their cycling accessibility in terms of both their ease of reaching destinations and their exposure to traffic stress (Kent and Karner 2019).

7 Conclusion

To conclude, the creation of (e-)cyclable cities holds the promise to transform urban mobility and promote greater transport safety, equity, public health, and more environmentally sustainable environments. However, achieving this requires going beyond a narrow focus on cycling infrastructure. Mainstreaming a culture of (e-)cycling depends on broader urban policies aligned with the principles of the 15-minute city—promoting compact urban form, diverse land uses, and close proximity to essential services. Only then can (e-)cycling become a practical and inclusive option for all. Finally, non-urban (e-)cycling raises ethical issues that only partially overlap with more traditional concerns in transportation ethics, bordering with questions about the ethics of tourism that need to be properly addressed.

References

Allen J, Farber S (2019) Sizing up transport poverty: a national scale accounting of low-income households suffering from inaccessibility in Canada, and what to do about it. Transp Policy 74:214–223

Ballo L, De Freitas LM, Meister A, Axhausen KW (2023) The E-Bike City as a radical shift toward zero-emission transport: sustainable? Equitable? Desirable? J Transp Geogr 111:103663

Berk T, Halvachizadeh S, Backup J et al (2022) Increased injury severity and hospitalization rates following crashes with e-bikes versus conventional bicycles: an observational cohort study from

a regional level II trauma center in Switzerland. Patient Saf Surg 16:11. https://doi.org/10.1186/s13037-022-00318-9
Berke A, Truitt W, Larson K (2024) Is access to public bike-share networks equitable? A multiyear spatial analysis across 5 US Cities. J Transp Geogr 114:103759
Björnberg KE, Hansson SO, Belin M-Å, Tingvall C (Eds) (2022) The vision zero handbook. In: Theory, technology and management for a zero casualty policy. Springer, Cham. https://doi.org/10.1007/978-3-030-76505-7
Brown C, Deka D, Jain A, Grover A, Xie Q (2019) Evaluating spatial equity in bike share systems. Rutgers University. https://doi.org/10.7282/t3-cs30-ad47
Byrne D (2021) Bicycle diaries. Faber & Faber, London
Castro A, Gaupp-Berghausen M, Dons E, Standaert A, Laeremans M, Clark A, Anaya-Boig E, Cole-Hunter T, Avila-Palencia I, Rojas-Rueda D, Nieuwenhuijsen M, Gerike R, Int Panis L, de Nazelle C, Brand A, Raser E, Kahlmeier S, Götschi T (2019) Physical activity of electric bicycle users compared to conventional bicycle users and non-cyclists: insights based on health and transport data from an online survey in seven European cities. Transp Res Interdisc Perspect 1(2019):100017. https://doi.org/10.1016/j.trip.2019.100017
Colville-Andersen M (2018) Copenhagenize. In: The definitive guide to global bicycle urbanism. Island Press, Washington-Covelo
Cunha I, Silva C (2023) Equity impacts of cycling: examining the spatial-social distribution of bicycle-related benefits. Int J Sustain Transp 17(6):573–591
Cunha I, Silva C, Büttner B, Toivonen T (2024) Pursuing cycling equity? A mixed-methods analysis of cycling plans in European cities. Transp Policy 145:237–246
de Hartog JJ, Boogaard H, Nijland H, Hoek G (2010) Do the health benefits of cycling outweigh the risks? Environ Health Perspect 118(8):1109–1116. https://doi.org/10.1289/ehp.0901747
Doran A, El-Geneidy A, Manaugh K (2021) The pursuit of cycling equity: a review of Canadian transport plans. J Transp Geogr 90:102927
Elvik R (2021) Cycling safety. In: Buehler R, Pucher J (eds) Cycling for sustainable cities. MIT Press, Cambridge, pp 58–79
Fuller D, Winters M (2017) Income inequalities in bike score and bicycling to work in Canada. J Transp Health 7:264–268
Garrard J, Rissel C, Bauman A (2012) Health benefits of cycling. In: Pucher J, Buehler (eds) City cycling. MIT Press, Cambridge, pp 31–55
Garrard J, Rissel C., Bauman, A., Giles-Corti, B. (2021). Cycling and health. In: Pucher J, Buehler (eds) Cycling for sustainable cities. MIT Press, Cambridge, pp 35–55
Gehlert T, Kröling S, Schreiber M, Schleinitz K (2018) Accidents analysis and comparison of bicycles and pedelecs. In: Framing the third cycling century. Bridging the gap between research and practice. German Environment Agency & European Cyclists' Federation, Dessau-Roßlau & Bruxelles, pp 77–87
Giuffrida N, Pilla F, Carroll P (2023) The social sustainability of cycling: assessing equity in the accessibility of bike-sharing services. J Transp Geogr 106:103490
Golub A, Hoffmann M, Lugo A, Sandoval G (2016) Bicycle justice and urban transformation. Routledge, New York
Gross I, Weiss DJ, Eliasi E, Bala M, Hashavya S (2018) E-bike–related trauma in children and adults. J Emerg Med 54(6):793–798
Heesch KC, Giles-Corti B, Turrell G (2015) Cycling for transport and recreation: associations with socio-economic position, environmental perceptions, and psychological disposition. Prev Med 63:29–35
Henderson J (2013) Street fight: the politics of mobility in San Francisco. University of Massachusetts Press
Huang Y, Jiang L, Chen H, Dave K, Parry T (2022) Comparative life cycle assessment of electric bikes for commuting in the UK. Transp Res Part D Transp Environ 105:103213. https://doi.org/10.1016/j.trd.2022.103213
Illich I (1974) Energy and equity. Marion Boyard, London

Jacobsen PL, Rutter H (2012) Cycling safety. In Pucher J, Buehler (eds) City cycling. MIT Press, Cambridge, pp 141–156

Kent M, Karner A (2019) Prioritizing low-stress and equitable bicycle networks using neighborhood-based accessibility measures. Int J Sustain Transp 13(2):100–110

Kroes TR, Poos HPAM, Buijze ALF, Rozema R, Lameijer H, Emous M (2025) Injury severity and traumatic brain injuries in youth and adolescents using conventional and electronic bicycles; a retrospective cohort study. Eur J Trauma Emerg Surg Official Publ Eur Trauma Soc 51(1):268. https://doi.org/10.1007/s00068-025-02941-4

Kroesen M (2017) To what extent do e-bikes substitute travel by other modes? Evidence from the Netherlands. Transp Res Part D Transp Environ 53:377–387. https://doi.org/10.1016/j.trd.2017.04.036

Marshall W (2024) Killed by a traffic engineer. In: Shattering the delusion that science underlies our transportation system. Island Press, Washington-Covelo

Martens K, Golub A, Hamre A (2021) Social justice and cycling. In: Buehler R, Pucher J (eds) Cycling for sustainable cities. MIT Press, Cambridge, pp 258–279

McNeil N, Dill J, MacArthur J, Broach J, Howland S (2017) Breaking barriers to bike share: insights on equity. National Institute for Transportation and Communities

Montgomery C (2013) Happy city: transforming our lives through urban design. Penguin UK

Moreno C (2015) The 15-minute city. In: A solution to saving our time & our planet. Wiley, Hoboken

Mueller N, Rojas-Rueda D, Cole-Hunter T, De Nazelle A, Dons E, Gerike R, Götschi T, Panis LI, Kahlmeier S, Nieuwenhuijsen M (2015) Health impact assessment of active transportation: a systematic review. Prev Med 76:103–114

Philips I, Anable J, Chatterton T (2022) E-bikes and their capability to reduce car CO_2 emissions. Transp Policy 116:11–23. https://doi.org/10.1016/j.tranpol.2021.11.019

PIARC—The World Road Association (2023) Biking for the future. In: A study of interurban cycling infrastructure and how to move forward. https://www.piarc.org/en/order-library/42702-en-Bikingfor%20the%20Future%20%E2%80%93%20Special%20Project

Piatkowski D (2024) Bicycle city. In: Riding the bike boom to a brighter future. Island Press, Washington-Covelo

Richardson BC (1995) Transportation ethics. Transp Q 49(2):117–126

Santilli G, Soldavini P (2019) Bikeconom. In: Viaggio nel mondo della pedalata. Egea, Milano

Santoni de Sio F (2021) The European commission report on ethics of connected and automated vehicles and the future of ethics of transportation. Ethics Inf Technol 23:713–726. https://doi.org/10.1007/s10676-021-09609-8

Schepers JP, Fishman E, den Hertog P, Klein Wolt K, Schwab AL (2014) The safety of electrically assisted bicycles compared to classic bicycles. Accid Anal Prev 73:174–180. https://doi.org/10.1016/j.aap.2014.09.010

Sundfør HB, Fyhri A, Bjørnarå HB (2020) E-bikes—good for public health? In: Nieuwenhuijsen MJ, Khreis H (eds) Advances in transportation and health, Elsevier, London, pp 251–266. https://doi.org/10.1016/B978-0-12-819136-1.00011-5

Tainio M, de Nazelle AJ, Götschi T, Kahlmeier S, Rojas-Rueda D, Nieuwenhuijsen MJ, de Sá TH, Kelly P, Woodcock J (2016) Can air pollution negate the health benefits of cycling and walking? Prev Med 87:233–236. https://doi.org/10.1016/j.ypmed.2016.02.002

Tortosa EV, Lovelace R, Heinen E, Mann RP (2021) Infrastructure is not enough. J Trans Land Use 14(1):693–714

Tortosa EV, Heinen E, Lovelace R (2022) Cycling and socioeconomic (dis) advantage. Adv Trans Policy Plann 10:211–233

Improving Accessibility in Urban Contexts Through E-Mobility

Matteo Colleoni and Rita Jankowski

Abstract E-bikes are often promoted as environmentally friendly alternatives to cars, offering an extended range compared to traditional bicycles, overcoming geographical challenges like distances, hills and slopes, serving as effective last-mile solutions when integrated with public transportation and enabling specifically elderly or people with reduced physical fitness to ride with less effort. They are more affordable than cars and do not require a driving license, making them accessible to a broader range of users. However, the potential of e-bikes to truly enhance urban accessibility for all is not guaranteed. This chapter examines e-bikes through the lens of inclusivity, exploring their role in creating a more equitable urban transport system. It discusses the necessary conditions for e-bikes to become a genuinely inclusive mobility option, such as access to safe and connected cycling infrastructure, secure bike parking, and seamless integration with public transport. The chapter also considers potential challenges and risks, including unequal access and safety concerns, that must be addressed to realize the full potential of e-bikes in improving urban mobility.

1 Introduction

Despite growing attention to the issue of environmental transition and EU guidelines for achieving climate neutrality, Italian cities continue to record high levels of polluting and climate-changing emissions. A significant proportion of these come from the transport sector, accounting for around a quarter of total EU greenhouse gas (GHG) emissions according to official statistics, 90% of which are due to road transport (EEA 2025; MIT 2025). The negative consequences are well known, rising average temperatures due to the greenhouse effect (GCP 2023), traffic congestion

M. Colleoni (✉) · R. Jankowski
University of Milano-Bicocca, Milan, Italy
e-mail: matteo.colleoni@unimib.it

R. Jankowski
e-mail: r.jankowski@campus.unimib.it

© The Author(s), under exclusive license to Springer Nature Switzerland AG 2026
R. Moscarelli et al. (eds.), *The Future of Slow Mobility*, Research for Development,
https://doi.org/10.1007/978-3-032-23660-9_3

and occupation of public space, with a consequent deterioration in the quality of the urban environment and people's quality of life.

Large fleets of polluting motor vehicles also have the effect of reducing mobility by public and active means (walking and cycling) and producing a modal split that is heavily dependent on cars. In Italy, for example, over the last 25 years, the number of vehicles on the road has increased by 34% (compared to a 5% increase in population) and has reached a total of approximately 55 million vehicles (of which almost 41 million are cars), equivalent to a motorization rate of 930 vehicles per 1000 inhabitants (ACI 2024).

This is a very large vehicle fleet, but also a polluting one, considering that about half of the cars belong to environmental classes lower than Euro 5 and that 84% of them are equipped with internal combustion engines (ICE, diesel or gasoline). Furthermore, if we consider the modal split, 67% of journeys are made by car or motorcycle, 8% by public transport, 21% on foot and only 4% by bicycle (ISFORT 2024).

Traffic congestion has negative consequences not only on the urban environment, but also on access to services and, more generally, on urban opportunities with alternative modes of transport, shaping urban space according to the circulation and parking needs of motor vehicles.

The issue of accessibility by foot and bicycle is particularly relevant in the contemporary debate on the city of proximity and, more specifically, on the 15-minute city. The concept of the 15-minute city, although not new, has seen renewed interest over the last decade alongside other models inspired by the city of proximity.

The reasons for this can be found in various factors that have converged in an unusual way in the combination of events over the last decade. The first concerns the consequences of widespread urbanization, which has led to increased land consumption and greenhouse gas emissions and a deterioration in the quality of the urban environment. The second factor concerns the COVID-19 pandemic, during which the urgent need for access to emergency services, particularly social and health services, had to contend with their scarcity or remoteness, especially in marginal areas.

These two factors, together with the measures adopted to resolve them and the difficulty of accessing services, were summarized in the discussion on the 15-minute city and, more generally, on urban accessibility to opportunities and social inclusion.

The essay summarizes the discussion, dedicating the first paragraph to the topic of proximity accessibility in contemporary cities and the second to cycling mobility and measures to improve active mobility in cities, with particular attention to the advantages offered by electric technologies applied to vehicles and infrastructure.

2 Proximity Accessibility in Contemporary Cities

The debate on the city of proximity is taking place at a time of extraordinary challenges for urban development, the most important of which concerns climate change and its consequences for urban ecosystems.

The last decade has been the warmest on record, the causes of which can be traced back to the increase in greenhouse gas (GHG) emissions due to human activities (Friedlingstein 2023). Climate change is associated with a second challenge for cities, namely strong population growth.

In 2024, the world population exceeded eight billion, with growth of over 30% since the beginning of the century and a projected increase that will bring it to 9.5 billion by 2050, 90% of which will be in developing countries and 20% over the age of 65. Population growth will mainly affect urban areas, which, despite covering a limited area, are home to over half of the world's population, a figure set to rise to 75% by 2050.

This anthropogenic pressure has the effect of increasing greenhouse gas concentrations and temperatures, with more pronounced consequences in urban environments and more negative effects on the most vulnerable populations, such as the sick and the elderly. In fact, it is mainly in cities that the relationship between resource consumption and greenhouse gas emissions is strongest.

As cities are where an increasing percentage of the population lives, works, and socializes, most greenhouse gas emissions are caused by buildings and transportation, sectors that have a significant impact on the quality of the urban environment (Tilche et al. 2023). The consequences of unsustainable housing and transport systems include not only high greenhouse gas (GHG) emissions and related damage to the health of inhabitants, but also traffic congestion, occupation and degradation of public spaces, noise pollution, and the risk of accidents, particularly for the most vulnerable road users.

These challenges justify the focus on local accessibility, particularly in the contemporary debate on the 15-minute city proposed by Carlos Moreno (2024) and other scholars (Zahaeer et al. 2022; Allam et al. 2022; Gall et al. 2024). Associated with the image of a pleasant, people-friendly city that is predominantly pedestrian and cycle-friendly, economically dynamic, environmentally friendly, socially inclusive, and culturally vibrant, it is presented as an alternative to urban development associated with excessive noise, pollution, long commutes, traffic congestion, poor quality of life, zoning, segregation, and inefficient use of time and space (Saliez 2024).

The proximity city model suggests a transformative planning theory that rethinks urban space by emphasizing the principle of proximity, according to which essential services must be accessible within a limited space–time threshold from homes by walking or cycling. This approach challenges the status quo of urban development by promoting a polycentric structure in which smaller activity centres reduce the need for car travel and generate communities. An example of functional spatial integration, the underlying assumption is that the optimal distribution of basic services leads to a reduction in the demand for motorized travel, which in turn frees up time and space that can be used to improve the urban experience.

Although the desire to save time has always been associated with transportation, the last century was characterized by supply-side solutions based on increasing the number, speed, and comfort of vehicles. The city of proximity, on the other hand, acts on the demand side, paying attention to the conditions of access of actors to services, the true objective of mobility, of which transport is merely an instrumental dimension, and to their equitable distribution throughout the territory.

The focus on accessibility, spatial equity, and proximity gives the city of proximity a strong connotation based on values that, although present in other urban planning theories, bring it closer to the arguments of scholars of transportation, social inclusion, and environmental justice (Lucas 2004; Sheller 2018). In particular, the emphasis on access conditions refers to Kaufmann et al.'s work on motility (2004), the set of socio-demographic, attitudinal, and material attributes that enable actors to access the services, places, and social networks in which they can meet their basic needs and participate in the social and economic life of the communities to which they belong (Kenyon et al. 2002).

The focus on spatial equity in the distribution of services and infrastructure is part of the debate on proximity services and the characteristics of urban space that promote accessibility. The ability to access essential services on foot or by bicycle, regardless of the time of day, which is simply an indication of an acceptable space–time threshold, raises the question of the criteria underlying the definition of a proximity service.

Already addressed by various authors, who share the belief that in contemporary societies the centrality of services refers to their ability to satisfy the set of needs that underpin the socio-economic and relational well-being of individuals (Colleoni 2024), in sectoral studies the theme has been interpreted in a spatial–temporal sense in terms of proximity services. According to Martens, these include services but also places of socialization where basic needs can be met, suggesting differentiated distribution policies according to the type of individual and modes of access tailored to their characteristics and skills (2017). The issue of skills is particularly relevant in urban societies in developed countries, where the high number of older people requires attention to be paid to measures to support their mobility and orientation in accessing services and places of interaction and social inclusion.

The latter aspect focuses on the second issue in the debate on spatial equity in the proximity city, relating to the characteristics of space that promote accessibility on foot and by bicycle. The debate on this topic can be summarized in the theory of the so-called "five Ds of public space" and in more recent theories on "proximity immobility." Originally proposed by Ewing and Cervero, the former summarizes in the five Ds of public space (high density and diversity of land use, design, accessibility of destinations, and short distance from public transport) the characteristics that public space must have to promote accessibility (2010).

Theories on proximity immobility, on the other hand, refer to the reorganization of urban space and the distribution of services based on the principle of reducing travel times and distances. The former is based on the Manifesto "plan more to travel less," according to which voluntary forms of limiting daily mobility could result from the implementation of integrated land-use and transport policies focused on improving physical and virtual accessibility through the proximity of opportunities and spatial and social networks (Banister 2005).

The same principle of ensuring proximity not through high mobility but through a combination of accessibility, virtuality, and liveability of space is found in the proximity planning proposed by Gehl (2011). Finally, some scholars believe it is necessary to explore the positive implications of a shift from a development paradigm based on globalization and high mobility to models centered on the principle of a local society inspired by slowness and localism (Ferreira et al. 2017). Reformulating Kaufmann's concept of 'motility', these researchers propose the concept of 'immotility', in which immobility, if reversible, is conceived as a form of capital in localist societies, where low levels of mobility are not perceived as the result of a constraint, but rather as a positive factor of closeness to the physical and social environment.

In conclusion, although difficult to achieve, it should be emphasized that the narrative of the highly accessible and close city, like other visions and utopias of the past and present, is creating a shared imaginary that aligns public and private actors around the common goal of improving the quality and well-being of cities in a historical phase characterized by unprecedented environmental and socio-economic challenges. Proposals on how to achieve this goal and the benefits it could bring are also contributing to the development of a common culture among public officials, designers, and citizens in general on what a good city is and how it should be designed (UN-Habitat 2022).

As part of the ASI (Avoid-Shift-Improve) approach to sustainable mobility policies, new electric mobility technologies applied to carriers and cycling networks can finally make a significant contribution to reducing fleets of internal combustion vehicles and greenhouse gas (GHG) emissions, as well as improving local accessibility to urban opportunities.

3 Electric Bikes and Urban Proximity

Particularly in urban areas, cycling is considered as a viable alternative to motorized vehicles, contributing to a cleaner, more sustainable, and liveable urban environment. The environmental benefits of cycling are well-known. According to estimations of UN Environment, switching from car to bicycle saves 150 g CO_2 per kilometre per person (UNEP 2019). Several studies demonstrate that both the direct costs of cycling infrastructure and the social costs considering health and environmental benefits are significantly lower for cycling compared to motorized vehicles (Sælensminde 2004; Gotschi 2011; Macmillan et al. 2014; Brey et al. 2017; Rich et al. 2021). In the concept of the proximity city, especially the 15-minute city, cycling is a fundamental pillar, as the model envisions that various urban nodes and essential services should be accessible within a 15-min walk or bike ride (Moreno 2019).

Taking the concept of the proximity city, we can see how different modes of transport enable access to different distances. While walking allows access to a radius of approximately 1.2 to 1.5 km in 15 min (assuming an average walking speed of around 5 km/h), cycling significantly extends this range. In Germany, the average cycling speed is approximately 14 km/h, although this can vary depending on factors such as age, topography, and urban context (Bikemap Blog 2021). Over a 15-min trip, this corresponds to a travel distance of about 3.5 km for cyclists. Electric bicycles further expand the 15-min travel radius. With pedal-assist motors, e-bike users tend to achieve higher average speeds than those on conventional bicycles. Schleinitz and colleagues found that riders of pedelecs (electric bikes with assistance up to 25 km/h) averaged around 17.4 km/h—roughly 2 km/h faster than traditional cyclists (Schleinitz et al. 2017). Other studies in Europe have reported average e-bike speeds ranging from 17 to 19 km/h in urban settings. At these speeds, an e-bike rider can cover approximately 4 to 4.75 km within 15 min, extending access far beyond what is typically possible with walking or conventional cycling.

Besides extending the proximity access range, conventional and electric bicycles generally have the potential to extend the functional boundaries of urban space. By requiring less physical effort and helping to overcome challenging topographical terrain, the average distance traveled by e-bike users is increasing. In urban spaces, with this extended proximity, e-bikes can serve as a serious alternative to cars for daily commuting, as these distances correspond to the average commuting range by car observed in European cities (European Commission. Directorate General for Mobility and Transport et al. 2022). When combined with public transportation, e-bikes can further expand the travel radius, improve first- and last-mile connectivity and shrink perceived distances in urban mobility. The Netherlands provides a leading example: roughly 47% of Dutch rail passengers reach the station by bicycle, illustrating how cycling extends the reach of transit well beyond walking range (Caletrio 2024).

As shown electric bicycles have the potential to reduce distance-related barriers in urban mobility by enabling higher speeds and requiring less physical effort than conventional bicycles allowing longer distances. However, accessibility must also be understood through the lens of capability, as introduced by Amartya Sen. The capability approach emphasizes that equal access to infrastructure does not necessarily result in equal real opportunities to use it, as individuals differ in their personal, social, and physical circumstances (Sen 1983). In this view, having access to a bicycle or cycling infrastructure does not automatically grant everyone the effective ability to reach destinations. What matters is not only the theoretical opportunity, but whether individuals have the actual capacity to convert that opportunity into meaningful use. For this reason, Sen argues that attention must be paid not only to available resources, but to people's ability to utilize them, so according to his capabilities concept. This is also proven by empirical analysis, where studies indicate that certain groups—such as women cycling (Tin et al. 2009; Green et al. 2010; Merom et al. 2010; de Sa et al. 2016; Bandeira et al. 2017; Aguilar-Farias et al. 2019), elderly and children (Goel et al. 2022)—remain underrepresented in. Among the barriers reported for these groups are physical fitness limitations and concerns around safety or infrastructure (Frater and Kingham 2018; Pearson 2023). For some of these reported barriers to cycling—such as physical exertion, limited fitness, or age-related constraints—e-bikes can serve as an alternative. For older adults and individuals with physical limitations, e-bikes can make cycling more manageable and less strenuous, enabling continued or renewed participation in active mobility. Considering people with disabilities, and considering that circumstances are very different, the bicycle could be a useful tool to enable independent movement. A survey among disabled cyclists done by the charity "Wheels for Wellbeing" in the UK found that 60% of the participants found cycling easier than walking and many use their cycle as a mobility aid (Wheels for Wellbeing 2021). Adapted e-bike models, such as tricycles and recumbent bikes with electric assistance can be an inclusive alternative here. These options support active aging, social participation, and physical activity, aligning with public health goals and promoting autonomy in daily life (Haustein and Siren 2015).

E-bikes can play an important role in supporting women's mobility, particularly in relation to care work and trip-chaining. Research has shown that women are more likely than men to engage in complex travel patterns that involve multiple stops for caregiving activities such as accompanying children, grocery shopping, or visiting relatives. As Wild et al. emphasize, e-bikes help alleviate common barriers to these activities—including physical exertion, time constraints, and the challenge of transporting children or household goods. For mothers in particular, e-bikes offer a practical solution to managing schedules by enabling commuting, caregiving, and errands to be combined into a single journey. Cargo e-bikes, in particular, were noted by study participants as tools that "democratise" cycling, making it helpful for heavier daily transport needs (Wild et al. 2021).

Beyond their pedal-assist function, modern e-bikes are often equipped with integrated digital communication technologies. These devices improve mobility and accessibility by enabling the transmission and reception of data related to routing and safety. Connectivity allows cyclists to remain digitally visible even in low-visibility conditions or adverse weather. Moreover, connected systems can support real-time navigation and safety alerts. In the event of an accident or hazardous situation, riders are able to take preventive measures or request assistance.

4 Requirements for Inclusive Cycling

The potential of e-bikes to enhance proximity, accessibility, and individual capability—and to serve as a genuine alternative to other modes of transport—is however inseparably linked to the quality of cycling infrastructure. To realise the benefits of e-bikes, infrastructure must support not only conventional cycling but also the specific needs associated with electric bikes. This includes the development of AAA cycling infrastructure—that is, infrastructure suitable for "All Ages and Abilities"—enabling all interested people to cycle safely and comfortably (National Association of City Transportation Officials 2017).

Such infrastructure should consist of protected and well-connected cycling routes, clearly separated from both motor vehicle traffic and pedestrian pathways. Adequate width is particularly important in the context of e-bikes, as their growing presence contributes to greater heterogeneity in cycling speeds. Sufficient space is needed to ensure safe overtaking and to accommodate a diverse range of users, including cargo bikes, tricycles, and other non-standard bicycles. Clear separation from pedestrian paths is especially important in the case of e-bikes, given their higher average speeds and potential for conflict in mixed-use spaces.

Such high-quality infrastructure is a prerequisite for inclusive cycling, as research indicates that fear of road traffic is one of the major concerns for people when it comes to cycling (Davies 1997; Fishman et al. 2012; Lawson et al. 2013; Sanders 2015; Pearson 2023).

Further infrastructure requirements include secure, weather-protected parking. Given the higher value of e-bikes, theft is a major concern for many users. The lack of safe parking facilities—particularly in residential areas and near transport hubs—can discourage people from purchasing or regularly using e-bikes. This issue is especially relevant for lower-income groups, who already express concern about bike theft in general and may be even more hesitant to risk investing in a more expensive e-bike without proper security infrastructure in place (Ledsham et al. n.d.; Christie et al. 2011; Vietinghoff 2021).

When talking about the inclusivity of e-bikes, it is also necessary to mention the issue of affordability, alongside their various potentials. While e-bikes can be a cost-effective alternative to car ownership, they are still significantly more expensive than conventional bicycles, making them less accessible to low-income groups. In addition to the high purchase cost, there are ongoing expenses such as maintenance, battery replacement, and theft protection.

Bike-sharing systems, when paired with appropriate infrastructure, can provide a practical alternative for urban mobility (International Energy Agency 2024). However, to be truly inclusive, these systems must extend beyond central urban areas and high-income neighbourhoods. For example, in Milan, the BikeMi bike-sharing system is available in areas with high socio-economic vulnerability. Yet, in practice, it tends to be more accessible to those who already have multiple mobility options, while underserved populations may still face barriers such as digital access, payment methods, or limited docking infrastructure in peripheral areas (Jankowski 2024).

5 Conclusions

In recent years, the environmental transition has gained growing attention across Europe, with the EU setting ambitious goals for climate neutrality. Nonetheless, many Italian cities continue to struggle with high levels of greenhouse gas emissions, particularly from the transport sector. Italy stands out for its exceptionally high rate of vehicle ownership and a mobility system heavily reliant on private cars. This entrenched car dependency not only contributes significantly to air pollution and traffic congestion but also hampers the growth of sustainable alternatives such as public transit and active modes of transport. As a result, urban spaces are often shaped more by the demands of car traffic than by the needs of people.

The proximity city models offer an alternative vision, aiming to restructure urban space around walking, cycling, and equitable access to essential services. At the core of the proximity city model is a planning approach grounded in the idea that basic services should be reachable within a limited space–time threshold from home, preferably on foot or by bicycle. This approach supports a polycentric urban structure that helps to reduce car dependence, encourages local living, and strengthens community life. Its focus on accessibility, spatial equity, and proximity closely aligns with key concerns in the fields of transport justice, environmental sustainability, and social inclusion.

E-bikes offer real potential to support more sustainable and accessible urban mobility. They make it easier to cover longer distances without a car, reduce the physical effort needed to cycle, and can help more people—including elderly people, women, and those with limited mobility—use bikes for everyday travel. When combined with public transport, they can also improve first- and last-mile connections, helping to reduce car dependency and making more parts of the city reachable without driving.

At the same time, their wider adoption depends on changes in how cities are planned. Good cycling infrastructure—safe, connected, and separated from traffic with adequate width—is essential if e-bikes are to become a real alternative for more people. Secure and weather-protected parking is also important, especially given the high cost of e-bikes and the concern about theft. Affordability more broadly remains a key barrier, and without targeted support or inclusive bike-sharing options, many people may still be left out.

References

ACI—Automobil club Italia (2024) Annuario statistico 2024

Aguilar-Farias N, Cortinez-O'Ryan A, Chandia-Poblete D, Heesch KC (2019) Prevalence and correlates of transport cycling in Chile. J Trans Health 14

Allam Z, Nieuwenhuijsen M, Chabaud D, Moreno C (2022) The 15-minute city offers a new framework for sustainability, liveability, and health. Lancet Planet Health 6(3):181–183

Bandeira ADS, da Silva KS, Del Duca GF, Zilch GR, de Oliveira ESA, de Barros MVG, Nahas MV (2017) Factors associated with bicycle use among Brazilian workers. Sport Sci Health 13(1):63–68

Banister D (2005) Unsustainable transport: city transport in the new century. Routledge, London

Bikemap Blog (2021) Deutscher fahrrad-index, bikemap blog, 17 Aug

Brey R, Castillo-Manzano JI, Castro-Nuno M, Lopez-Valpuesta L, Marchena-Gomez M, Sanchez-Braza A (2017) Urban land for cycling promotion: cost-benefit analysis. Land Use Policy 63:130–139

Caletrio J (2024) The combined use of bicycles and trains in the Netherlands. Mobile Lives Forum

Christie N, Kimberlee R, Towner E, Rodgers S, Ward H, Sleney J, Lyons R (2011) Children and cycling in disadvantaged areas. J Transp Geogr 19(4):943–949

CNAPPC—Consiglio nazionale degli Architetti, Pianificatori, Paesaggisti e Conservatori (2024) Italy in proximity. In: The future of urban and territorial planning. Lettera Ventidue Edizioni, Siracusa

Colleoni M (2024) Mobilité, immobilité et risque d'exclusion sociale. In: Kaufmann V et al (eds) Les nouvelles proximités, tome 1, FYP Éditions, pp 45–62

Davies DG (1997) Attitudes to cycling: a qualitative study and conceptual framework. Transport Research Laboratory, Crowthorne
de Sa TH, Pereira RHM, Duran AC, Monteiro CA (2016) Active transportation in Brazil, vol 50. Revista de Saude Publica
European Commission et al. (2022) Study on new mobility patterns in European Cities: executive summary. Publications Office of the EU, Luxembourg
European Environment Agency (EEA) (2025) Transport and mobility 2025. https://www.eea.europa.eu/en/topics/in-depth/transport-and-mobility
Ewing R, Cervero R (2010) Travel and the built environment: a meta-analysis. J Am Plann Assoc 76(3):265–294
Ferreira A, Bertolini L, Næss P (2017) Immotility as resilience? Appl Mobilities 2(1):16–31
Fishman E, Washington S, Haworth N (2012) Fear of bicycle riding in Australia. J Australas Coll Road Saf 23, pp 19–27
Frater J, Kingham S (2018) Gender equity and adolescent cycling. J Trans Geogr 71, pp 130–138
Friedlingstein P (ed) (2023) Global carbon budget 2023. In: Earth system science data, vol 15, no 12, pp 5301–5369
Gall C, Gwiazdzinski L, Kaufmann V, Torre A (2024) Les nouvelles proximités, tome 1, FYP Éditions
GCP—Global Carbon Project (2023) Global carbon budget report 2023
Gehl J (2011) Life between buildings. Island Press, Washington D.C
Goel R, Goodman A, Aldred R et al (2022) Cycling behaviour in 17 countries. Transp Rev 42(1):58–81
Gotschi T (2011) Costs and benefits of bicycling investments in Portland. J Phys Act Health 8(S1):S49–S58
Green J, Steinbach R, Datta J, Edwards P (2010) Cycling in London, smarter travel unit, transport for London
Haustein S, Siren A (2015) Older people's mobility. Transp Rev 35(4):466–487
International Energy Agency (2024) World energy outlook 2024
Isfort (2024) Ventunesimo rapporto sulla mobilità degli italiani, Roma
Jankowski R (2024) Cycling justice: inequalities in Milan and Berlin, University of Milano-Bicocca
Kaufmann V, Bergman MM, Joye D (2004) Motility: mobility as capital. Int J Urban Reg Res 28(4):745–756
Kenyon S, Lyons G, Rafferty J (2002) Transport and social exclusion. J Transp Geogr 10(3):207–219
Lawson AR, Pakrashi V, Ghosh B, Szeto WY (2013) Cyclist safety perception in Dublin. Accid Anal Prev 50:499–511
Ledsham T, Zhang YX, Farber S, Hess P, (n.d.) Cycling in low-income suburbs. Int J Sustain Transp
Lucas K (2004) Running on empty. In: Transport, social exclusion and environmental justice. Policy Press, Bristol
Macmillan A, Connor J, Witten K et al (2014) Commuter bicycling: costs and benefits. Environ Health Perspect 122(4):335–344
Martens K (2017) Transport justice. In: Designing fair transportation systems. Routledge, Abingdon
Merom D, van der Ploeg HP, Corpuz G, Bauman AE (2010) Public health and travel surveys. Am J Prev Med 39(2):113–121
MIT—Ministero delle Infrastrutture e dei Trasporti (2025) Osservatorio sulle tendenze della mobilità. III trimestre
Moreno C (2024) The 15-minute city: a solution to saving our time and our planet. Wiley, London
Moreno C (2019) The 15-minute city: for a new chrono-urbanism. Carlos Moreno Blog, 29 Dec
National Association of City Transportation Officials (2017) Designing for all ages & abilities, NACTO
Pearson L (2023) Gendered barriers to cycling. 3CR Community Radio
Rich J, Jensen AF, Pilegaard N, Hallberg M (2021) Bicycle infrastructure and e-bikes. Case Stud Trans Policy 9(2):608–615

Sælensminde K (2004) Cost–benefit analyses of walking and cycling networks. Transp Res Part A 38(8):593–606
Saliez F (2024) Sustainability & governance in the city of proximity. In: CNAPPC, Italy in proximity, pp 123–139
Sanders RL (2015) Perceived traffic risk for cyclists. Accid Anal Prev 75:26–34
Schleinitz K, Petzoldt T, Franke-Bartholdt L, Krems J, Gehlert T (2017) Cycling speed of e-bikes vs conventional bikes. Saf Sci 92:290–297
Sen A (1983) Poor, relatively speaking. Oxf Econ Pap 35(2):153–169
Sheller M (2018) Mobility justice: the politics of movement in an age of extremes. Verso Books
Tilche A, Basile FL, Torsello M (2023) Le città a impatto climatico zero. Il Mulino, Bologna
Tin TS, Woodward A, Thornley S, Ameratunga S (2009) Cycling and walking in New Zealand. Int J Behav Nutr Phys Act 6(1):64
UNEP (2019) Cycling, the Better Mode Of Transport. United Nations Environment Programme
UN-Habitat (2022) Let's embrace the 15-minute city. Concepts, World Cities Report
Vietinghoff C (2021) Barriers to cycling in a French city. J Trans Geogr 91
Wheels for Wellbeing (2021) Disability & Cycling. Report of 2021 National Survey Results
Wild K, Woodward A, Shaw C (2021) Gender and the e-bike. Active Travel Stud 1(1)
Yunus M (2017) A world of three zeros. PublicAffairs
Zahaeer A, Chabaud D, Gall C, Pratlong F, Moreno C (2022) Resilient and sustainable cities. Elsevier

Planning for Urban Slow E-Mobility: A Synthetic and Systematic Review of Social, Infrastructural, and Safety Challenges

Martina Carra and Michele Pezzagno

Abstract The rapid spread of soft and electric-powered micro personal mobility vehicles (S&ePMVs) such as e-bikes and e-scooters is redefining urban mobility debates. Their integration in the urban context remains ambiguous and a topic of growing interest. Several studies and reviews focused on different topic-related, e.g., micro-mobility and built environment relationship, safe bikeway infrastructure design, or e-bike and e-scooter level-of-service. However, few studies investigated a unifying approach to planning, infrastructural, and social issues related to their integration. Therefore, this paper aims to identify the role of S&ePMVs within transport and urban planning, and how they are integrated into planning, infrastructural, and social issues, as well as opportunities and challenges therein. The synthetic and systematic literature review is based on PRISMA methodology and examined 49 articles from 2013 to 2025. As a result, three analytical domains are examined: (i) social challenges, (ii) planning strategies, and (iii) safety concerns and emerging conflicts. Results revealed persistent tensions, e.g., disparities related to gender and age, uneven infrastructural continuity and street space reallocation, and heightened safety risks stemming from poor design. Therefore, while S&ePMVs promise to favour sustainable mobility and reduce car dependency, their diffusion also risks deepening inequalities and generating new conflicts. Finally, results emphasised policy implications, emphasising more flexible and forward-thinking strategies to integrate S&ePMVs. This study provides urban planners and policymakers with evidence-based guidelines for integrating S&ePMVs into transport systems, balancing sustainability goals with equity and safety concerns.

M. Carra (✉) · M. Pezzagno
DICATAM, Università degli Studi di Brescia, Brescia, Italy
e-mail: martina.carra@unibs.it

M. Pezzagno
e-mail: michele.pezzagno@unibs.it

© The Author(s), under exclusive license to Springer Nature Switzerland AG 2026

R. Moscarelli et al. (eds.), *The Future of Slow Mobility*, Research for Development,
https://doi.org/10.1007/978-3-032-23660-9_4

1 Introduction

Recently, cities have been facing increasingly complex challenges due to negative outcomes of car-centred mobility models and the introduction of innovative mobility opportunities. Factors such as climate change, air pollution, traffic congestion, and the growing demand for sustainable mobility call for a transformation of urban transport systems and spaces connected to them. In this context, soft and electric-personal mobility vehicles (S&ePMVs)[1] i.e e-bikes, pedelecs, and e-scooters, have assumed a relevant role (Boglietti et al. 2021; Bretones and Marquet 2022). The global S&ePMVs market has been growing in the past few years and is currently an option for urban mobility as an alternative to private cars and complementing public transport (Sabbaghian et al. 2023; Lin et al. 2017; Fan and Harper 2022). However, this growth introduced new challenges of compatibility (e.g., Gössling 2020).

Consequently, a rapidly expanding strand of literature has focused on S&ePMVs from several perspectives, e.g., user, management, and operational. Specifically, studies have analysed how S&ePMVs generate new spatial and aspatial forms of coexistence and cohabitation. While they can enhance opportunities for inclusion and independence, they expose limits to perceived safety, gender equality, and spatial justice (e.g., Cubells et al. 2023a; Fyhri et al. 2017). Parallelly, several studies have addressed the systemic re-planning of infrastructures, highlighting relationship patterns of human activity to the built environment, and concept and methods to redistribute (or reallocate) "road" space (e.g., Zhang et al. 2023; Ballo et al. 2023). Finally, some authors have deepened the topic related to speed profiles, inadequate infrastructures, and unpredictable behaviours, which generate both objective and perceived risks, and strongly affect the propensity to use S&ePMVs and the broader social legitimacy of active mobility users (e.g., Porojkow and Lißner 2024; Kazemzadeh 2025).

The debate is open and increasingly attracting attention from both scholars and practitioners. Therefore, this study provides a critical review of the literature (2013–2025) on three analytical domains of S&ePMVs: social, planning, and safety issues. The aim is to highlight unresolved tensions and emerging solutions, thereby offering a reference framework of knowledge to support decision-makers and urban/transport planners.

The paper is structured as follows: Sect. 2 describes the methodology adopted; Sect. 3 presents the results; Sect. 4 discusses the findings, concluding with a proposal of a set of guidelines.

---

[1] The literature offers multiple definitions of S&ePMVs, yet there is general consensus in delimiting the category to lightweight vehicles, typically under 35 kg, with maximum speeds of around 25 km/h, powered by small electric motors and suitable for short-distance travel. These include primarily e-bikes (pure, pedelec, and hybrid) and e-scooters, while other devices such as segways, hoverboards, and monowheels remain marginal and are therefore not considered in this paper. S&ePMVs may be operated either through shared services (docked or dockless) or in private ownership.

2 Methodology

The review is based on the Preferred Reporting Items for Systematic reviews and Meta-Analyses (PRISMA) framework, structured into four phases, i.e., identification, screening, eligibility, and inclusion (Moher et al. 2009). The research question focused on the role of S&ePMVs within transport and urban planning, and how they are integrated into planning, infrastructural, social, and governance issues, as well as opportunities and challenges therein.

The identification phase consists of a comprehensive search in the Scopus database, using a keyword string covering vehicles types ("e-mobility", "micro mobility", "electric micro vehicles", "e-scooter", "e-bike", "ebike", "ecargo", "e-cargo", "personal mobility vehicle", "S&ePMVs ") and urban dimensions ("urban planning", "transport planning", "infrastructure", "shared urban spaces", "safety", "accessibility", "policy", "conflict", "innovation"). Keywords are combined with boolean (AND, OR) and truncation operators to form a precise and adaptable search string. The analysis resulted in a total of 2,385 records. During screening, irrelevant disciplinary fields (e.g., medicine, physics, chemistry; n = 332) and non-English contributions (n = 18) were excluded. Further filters restricted the dataset to journal articles/reviews (n = 419). Next, titles (n = 1,231) and abstracts (n = 212) were analysed for relevance, resulting in 273 records. The eligibility phase excluded additional studies marginal to the research focus (n = 214). To ensure conceptual clarity, shared mobility studies were excluded given the heterogeneity introduced by business models, operating rules, and redistribution dynamics. Therefore, studies not addressing individually owned S&ePMVs are excluded. Those focused on non-urban areas or primarily focused on accessibility to public transport or travel behaviour unrelated to infrastructure or planning, since it is not the aim of the present study, as well as those centred on pollution, batteries, or energy storage. Obviously, unavailable texts were eliminated.

Lastly, 49 articles between 2013 and 2025 resulted in the inclusion phase and were then analysed, highlighting their exponential interest (Fig. 1).

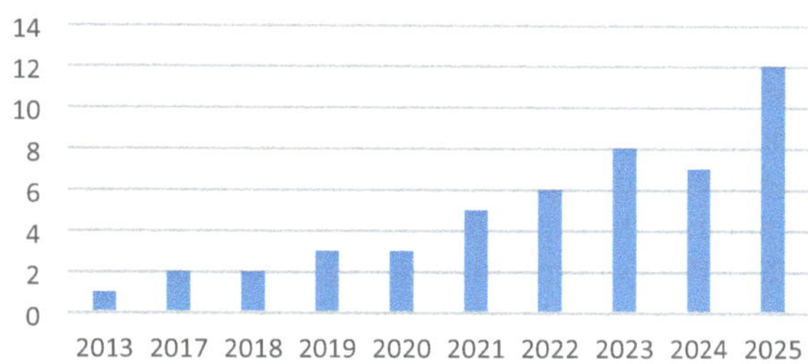

Fig. 1 Number of publications reviewed per year: 2013–2025

3 Results

The literature identifies three recurrent and critical analytical domains that constitute the focus of this review: social challenges, planning strategies and infrastructure provision, and safety concerns and emerging conflicts.

3.1 *Social Challenges*

Diffusion of S&ePMVs is not only a technological shift but also concerns social questions related to equity, safety, and the cohabitation of public space (Rainieri et al., 2024). Users are constantly required to negotiate their right to public space, sharing roads and sidewalks with cars and pedestrians. These interactions create conflicts: in car lanes, S&ePMV users feel vulnerable and disregarded; on sidewalks, their increased speed generates insecurity amongst pedestrians (Cubells et al., 2023a). While these tensions are intrinsic to the very nature of shared urban space, they largely depend on how co-existence is ruled and managed. These dynamics are also strongly influenced by people's behaviour, shaped by demographic and psychosocial factors that intervene in decision-making processes and in the formation of preferences (Bretones & Marquet, 2022).

Acceptance is complex and far from uniform: gender, age, socio-economic conditions, and individual perceptions shape patterns of access and everyday use. Two main demographic clusters emerge from the literature.

The first relates to gender dynamics. Several studies document the persistent gap between men and women in the use of S&ePMVs (Moinse 2025). Women are less likely to use these vehicles (particularly e-scooters) and when they do, they tend to restrict their trips to routes with dedicated infrastructure that is physically separated from motorised traffic (Cubells et al. 2023a). This imbalance reflects both structural and cultural factors. For instance, in countries such as the Netherlands, Denmark, and Germany, the development of continuous and safe cycling routes has resulted in a fairer distribution of use. Therefore, the most significant barriers are related to traffic protection and vulnerability to crash exposure (Moise 2025). Other obstacles concern greater visibility in public space while riding: more than one-third of women report having experienced at least one episode of harassment while using an e-bike or e-scooter (Adorean et al. 2025). As a result, many develop avoidance strategies such as not travelling at night, avoiding isolated or poorly lit routes, or limiting the use of S&ePMVs to specific contexts. Recent research suggests that S&ePMV adoption tends to exacerbate such inequalities, reinforcing the same gender divide already observed in traditional cycling.

The second cluster concerns older adults, who often withdraw from active mobility due to physical limitations or safety concerns. In this regard, S&ePMVs, especially e-bikes, can represent an opportunity, enabling them to maintain everyday independence and social networks even in later life (Leger et al. 2019; Van Cauwenberg et al. 2018). Assisted pedalling reduces effort and expands the range of feasible slopes and distances, without eliminating the health benefits of active mobility. In suburban areas, where public transportation provision is limited, S&ePMVs are often a possible resource in the prevention of social isolation. However, several obstacles persist, such as perceived insecurity at intersections, difficulties inherent in the dynamic encumbrance of sidewalks, the scarcity of dedicated and segregated cycle infrastructures, and the risk of single-vehicle falls. These findings highlight the need for more inclusive infrastructure design inspired by the principle of "All Ages and Abilities", especially considering that this demographic group is steadily growing (Leger et al. 2019).

Socio-psychological factors also play a critical role. People adopt S&ePMVs both for functional (i.e., time and cost savings, convenience) and symbolic (i.e., well-being, independence, social status, enjoyment, environmental sensitivity) motivations (Bretones and Marquet 2022; Rérat 2021). Nevertheless, perceived insecurity, a lack of dedicated infrastructure, the risk of theft, and high costs remain widespread barriers (Fyhri et al. 2017; Lin et al. 2017). Expectations further shape satisfaction levels: longitudinal studies in the Netherlands show that although satisfaction with e-bike trips remains high, it is often lower than initial expectations, pointing to a "focusing illusion" effect (de Kruijf et al. 2019). Similarly, Khavarian et al. (2024) observed that the expectations of users about travel time and convenience did not always match their experience, primarily due to infrastructure discontinuities that, in most cases, forced them to choose less optimal routes.

Finally, the question of spatial justice also emerges. Case studies in San Francisco and Montréal reveal how infrastructures and services often favour central districts over peripheral and less affluent areas, which remain dependent on private cars. This uneven provision shows how, when left to market dynamics alone, S&ePMVs may exacerbate territorial disparities and reinforce power imbalances between central urban areas and their hinterlands (Moran 2021; Romm et al. 2025).

In summary, the transition toward S&ePMVs is not a neutral process. It implies political and cultural choices that determine who will be able to obtain sustainable mobility and who will probably be denied.

3.2 Planning Strategies and Infrastructure Provision

Infrastructure planning constitutes a significant challenge. While initially, S&ePMVs were often interpreted as a spontaneous innovation, driven by commercial experimentation or local initiatives, their growing diffusion and heterogeneity today require a systemic rethinking of infrastructures. Three themes stand out as particularly relevant: the influence of the built environment, the development of dedicated networks and infrastructures, and the redistribution of public space in favour of S&ePMVs.

The literature examining the impact of the built environment on S&ePMV ownership and use has relied on the "D" framework (i.e., density, diversity, and design, later extended to accessibility, distance to public transport, demand management, and demographics) as an interpretive frame to explain the relationships between urban form and modal choice (Zhang et al. 2013, 2023; Zhao et al. 2018; Jin et al. 2023). Several studies had similar findings about the influence of residential density, functional mix, and public transport quality on S&ePMVs adoption. However, regional variations were observed. Specifically, Ding et al. (2019) have shown that these dynamics are non-linear, i.e., moderate density tends to foster it, while extremely high levels might discourage it due to competition with walking and public transport. In long distances and low-quality transit services, S&ePMVs often emerge as alternatives to cars, particularly for low- and middle-income populations. Socio-demographic factors also modulate the propensity to adopt them (Cubells et al. 2023b; Yu et al. 2022). Longitudinal research further confirms that the effects of the built environment consolidate over time: living in high-density areas or in contexts with good accessibility to destinations may initially reduce the use of S&ePMVs, but in the long run, induce a positive adaptation (Zhang et al. 2025).

Based on these variables, a second stream of studies has focused on methodological tools for planning dedicated networks and infrastructures to support decision-making. GIS-based and multicriteria approaches have been developed to prioritise infrastructure investments by considering demand, safety, proximity to points of interest, and crash data (Fazio et al. 2021). Demand has been estimated employing floating car data and predictive models to count the percentage of modal shift from cars to S&ePMVs. For instance, Comi and Polimeni (2024) have estimated a value of about 22% in medium-sized cities. Data-driven models have also integrated cycling crash data to demand flows to balance safety and territorial coverage objectives (Folco et al. 2022). Approaches and techniques such as space syntax and GPS use have enabled the analysis of route choice dynamics, incorporating network spatial structure into planning models (Schumann et al. 2025). These tools allow the transition from localised interventions to systemic network planning that accounts for both urban morphology and user routing preferences. However, the expansion of cycling networks tends to generate only marginal increases in already developed contexts, such as in Denmark, whereas it proves more significant in countries with a less consolidated cycling culture (Hallberg et al. 2021).

On the infrastructural side, quality is a decisive factor. Users tend to prefer continuous and extensive cycleways, physically separated from vehicular traffic, even if it requires a detour from the shortest route (Van den Steen et al. 2022; Meister et al. 2023). E-bikes tolerate longer distances and higher gradients. E-scooters are sensitive to pavement conditions, traffic lights, and pedestrian crowding (Cao 2025). Suitable lighting, urban furniture, and tree cover amplify attractiveness in both cases (Hu et al. 2024). The results indicated the need for synthetic standards of infrastructure with regard to the specifics of each S&ePMVs type.

A third group of studies focused on the more "radical" re-allocation of road space in favour of S&ePMVs. Ballo et al. (2023) proposed the "E-Bike City" as a new idea/starting point for transport policy discussions. Therefore, the authors outlined a research agenda for testing the effects of the re-allocation of car lanes into spaces for bicycles, pedestrians, and S&ePMVs without undermining the overall functionality of the transport system. Subsequent studies have formalised automated and non-automated processes for reallocating space at the network level, showing that it is possible to quadruple cycling capacity while maintaining adequate services for public transport and cars (Ballo et al. 2024; Romm et al. 2025; Zhu et al. 2025). Romm et al. (2025) further showed that even in scenarios of heavy investment and large-scale reallocation of infrastructures to S&ePMVs, vehicular traffic still retains the majority of roadway capacity. In such contexts, redistribution may also acquire a dimension of social justice, improving accessibility for those without private cars, especially in low-density suburban areas (Zhu et al. 2025). However, in high-density urban areas, the reduction of car spaces may disproportionately affect certain social groups, requiring compensatory policies to prevent exacerbating existing inequalities.

In summary, planning for S&ePMVs emerges as a multidimensional process that reflects broader tensions between incremental adaptation and radical redesign, and underscores that infrastructure provision is not only a mobility issue but also a spatial and environmental imperative.

3.3 *Safety Concerns and Emerging Conflicts*

Safety is one of the central issues in the diffusion of S&ePMVs: the introduction of new vehicle dynamics on infrastructures that are often inadequate can generate risks along network links and at intersections. Unlike traditional bicycles, S&ePMVs exhibit distinctive riding behaviours, such as rapid acceleration and varied speed profiles, which further complicate their coexistence with other transportation modes (Kazemzadeh and Ronchi 2021; Cai et al., 2020).

The issue of coexistence in shared spaces is particularly critical: sidewalks and non-motorised lanes are often merged into mixed-use environments. Mitropoulos et al. (2023) report that in more than half of the studies analysed, e-scooters were found to share the road with other users, generating tensions with both pedestrians and cyclists. Due to undersized and insufficiently designed infrastructure, a high frequency of overtaking manoeuvres and interactions forces S&ePMV users to occupy car lanes, thereby increasing the risk of collisions or adopting unpredictable behaviours (Porojkow and Lißner 2024). Trajectories and GPS data analyses also proved that frequent deviations are often related to infrastructures that are not continuous (Zuniga-Garcia et al. 2021; Cubells et al. 2023a; Porojkow and Lißner 2024; Mitropoulos et al. 2023). At the same time, the flexibility and unpredictability of pedestrian movements—the most vulnerable road users—often compel S&ePMV riders to perform sudden evasive manoeuvres, resulting in instability, difficulties in vehicle control, and increased risk of collisions with pedestrians or other obstacles (Sim and Cho 2025).

Many variables, such as the ratio of pedestrian and S&ePMVs flows and gender differences, influence the severity of conflicts. Homogeneous flows tend to reduce risks due to their more uniform speeds, whereas high heterogeneity among users increases the frequency of overtaking and collision probability (Miao et al. 2025).

The probability increases and gets worse with deviant behaviours, i.e., wrong-way riding or impeding lawful lane use due to illegal parking. These findings indicate the fragility of current strategies for managing shared spaces (e.g., Khavarian et al. 2024).

Speed is a crucial determinant. Research on riding patterns in Switzerland has revealed that e-bikes ride faster than conventional bicycles and s-pedelecs frequently exceed the 23 km/h limit, also within zones with a speed range of 20–30 km/h (Maurer et al. 2025). These discrepancies translate into slower reaction times and increased risk exposure. Infrastructure characteristics further affect behaviour: painted cycle lanes on roadways are associated with higher speeds, while segregated cycle tracks tend to induce slower riding. Speed-limiting regulations also have an influence on modal choices: stricter limits often push S&ePMV users onto sidewalks, increasing interactions with pedestrians, while higher limits favour their use of road lanes (Cicchino et al. 2023).

Infrastructure conditions are also critical. Poor pavement quality, fixed obstacles, and illegal parking are recurring causes of crashes and falls, particularly for e-scooter riders (Kazemzadeh and Sprei 2022; Kazemzadeh and Ronchi 2021). Experimental studies have shown that instability increases with speed and that irregular surfaces or height differences significantly amplify the risk of overturning, suggesting the need for stricter regulations and continuous surface maintenance (Kwon et al. 2024).

Finally, intersections have been analysed with respect to user behaviour near traffic lights. Cai et al. (2020) proved that S&ePMV users are more likely than traditional cyclists to cross at yellow or red lights, leveraging the difference in acceleration and speed. Similarly, van Lierop et al. (2020) on wayfinding signage on cycle highways, indicating that customised, conspicuous signs positioned in advance of decision points enhanced feelings of safety and reduced trajectory errors. These findings underscore the role of wayfinding systems and integrated design solutions that increase legibility and provide adequate reaction times (Kazemzadeh 2025).

4 Discussion and Conclusions

The results show that the diffusion of S&ePMVs represents a significant challenge for urban and transport planning. The evidence gives an ambivalent picture, suspended between promises and persistent challenges. Daily, users negotiate their right to public space (and, more broadly, to the city) in contexts marked by conflict, vulnerability, and asymmetries of power, while cities struggle to provide adequate infrastructures. These dynamics converge in highlighting a multidimensional issue characterised by substantial governance deficits.

From a social perspective, S&ePMVs raise questions of equity and spatial justice. Infrastructures concentrated in central and affluent neighbourhoods amplify exclusion and perceptions of insecurity and marginalisation (Moran 2021; Zhu et al. 2025). In peri-urban areas, which are often car-dependent and poorly served by public transport, the absence of dedicated facilities further accentuates the divide with urban cores. In this sense, policies cannot simply promote device adoption; instead, they must ask who is included and who remains excluded from new forms of mobility, due to both endogenous and exogenous factors (Adorean et al. 2025; Cubells et al. 2023a).

Studies focused on planning strategies reveal the tension between incremental interventions and transformative visions. While the built environment and network quality remain key determinants (Zhang et al. 2023), to enable permanent user retention (Khavarian et al. 2024), a one-size-fits-all approach is not sufficient. Instead, infrastructural continuity, diversified standards, and public space redistribution turn into conditions to scale up the phenomenon. Radical proposals, such as the "E-Bike City," highlighted that the challenge is both political and technical: how much can car space be reduced to increase active mobility? What regulatory regimes can resolve conflicts and favour co-existence (Ballo et al. 2024; Romm et al. 2025)?

Parallelly, evidence in the safety analytical domain reveals that deceleration or enforcing individual responsibility (e.g., helmets, age restrictions) alone is insufficient. Poorly legible mixed-use contexts, congested sidewalks, and critical intersections transform the promise of sustainability into new sources of risk (Cai et al. 2020; Sim and Cho 2025). A centre–periphery dichotomy emerges: in cities, conflicts are primarily caused by space over-saturation, while in suburbs, risks are caused by the lack of dedicated infrastructures and having to share carriageways with motorized transport. Safety here is not just to be considered a technical condition but also as being within a subjective dimension from which legitimacy and political response are derived (Mitropoulos et al. 2023).

Overall, these three domains underscore the role of governance. Disjointed and fragmented approaches, conflicting or overly directive rules, and wielding authoritative instruments in the lack of a precise and integrated vision can produce a policy vacuum that neither effectively protects users nor creates an actual paradigm shift (Curtis et al. 2025). Moreover, cost–benefit analyses show that although significant social and environmental returns exist, the additional (and high) safety-related costs require recalibration of strategies (Rich et al. 2021).

In light of these findings, several guidelines emerge that should be integrated into urban planning frameworks as operational suggestions for policymakers (Table 1).

In conclusion, this review aims to shed light on both the opportunities and critical issues associated with S&ePMVs. However, it remains limited to the longitudinal and comparative studies selected, thus pointing to the need for more systematic, multi-level analyses in the future, with particular emphasis on the ex-post evaluation of infrastructural changes.

Table 1 Policy implications and operational actions

Domain	Guidelines	Actions
Equity and inclusion	Integrate S&ePMVs into universal accessibility strategies (both physical and social), ensuring they do not reproduce or exacerbate existing inequalities	• Provide subsidies or incentives for vulnerable groups • Ensure equitable distribution of infrastructures, with particular focus on peripheral and peri-urban areas, and zones with poor public transport accessibility • Incorporate socio-psychological factors into campaigns and policies • Implement targeted awareness-raising campaigns
Infrastructure	Implement a dedicated cycle network, with differentiated standards and compatibility for different types of S-&PMVs	• Deliver a continuous, legible, and coherent system; • Prioritise physically protected and spatially consistent facilities; • Reclaim car space for pedestrians and S&ePMVs, avoiding "fallback" options on sidewalks • Map and classify the cycling network (e.g., width, continuity, design speed), identifying critical gaps (leveraging opportunities from machine learning techniques) • Employ technology and big data for monitoring and analysis • Enable multimodal integration with safe parking and micro-hubs for interchange

(continued)

Table 1 (continued)

Domain	Guidelines	Actions
Safety	Combine speed regulations and individual obligations with the redesign of critical nodes and clear separation of pedestrians and S&ePMVs	• Adopt technological monitoring tools • Harmonise regulations across jurisdictions (e.g., speed limits, parking, equipment standards) Address micro-structural road factors (geometry, visibility, pavement quality) • Redesign intersections with protected signal phases and set-back crossings • Introduce dynamic speed management on shared routes (e.g., GPS-based limitations) • Monitor usage and crash data by device type, enabling evidence-based adaptation
Regulation and governance	Overcome fragmentation and policy vacuums through national standards, geofencing, and adaptive tools	• Involve private actors and local communities, avoiding full delegation of management to market forces • Avoid punitive regulations unsupported by adequate infrastructure • Assign dedicated budgets for maintenance and ensure annual service provision (not limited to general roadway funds)

Acknowledgment This work was supported by (i) the Università degli Studi di Brescia & MUR within the Grant "CUP: D75F21002920001" PON R&I 2014–2020 (FSE REACT-EU) and (ii) the MOST – Sustainable Mobility National Research centre and received funding from the European Union Next- GenerationEU (PIANO NAZIONALE DI RIPRESA E RESILIENZA (PNRR) – MISSIONE 4 COMPONENTE 2, INVESTIMENTO 1.4 – D.D. 1033 17/06/2022, CN00000023), Spoke 5 "Light Vehicle and Active Mobility". This manuscript reflects only the authors' views and opinions, neither the European Union nor the European Commission can be considered responsible for them.

CRediT authorship contribution statement Conceptualization: M.P. & M.C.; Formal analysis: M.C.; Investigation: M.C.; Methodology: M.C.; Supervision: M.P.; Writing-Original Draft: M.C.; Writing–review & editing: M.P.

Declaration of generative AI and AI-assisted technologies During the preparation of this work, the authors used ChatGPT (OpenAI) to improve English language and readability. After using this tool, the authors reviewed and edited the content as needed and took full responsibility for the publication's content.

References

Adorean EC, Nofre J, García-Juan L, Moura F (2025) Perceptions of safety and security among e-bike and e-scooter users in Iberian capital cities: implications for urban mobility planning. Transport Res F Traffic Psychol Behav 114:977–1005. https://doi.org/10.1016/j.trf.2025.07.009

Barabino B (2018) Automatic recognition of "low-quality" vehicles and bus stops in bus services. Public Transport 10(2):257–289. https://doi.org/10.1007/s12469-018-0180-8

Boglietti S, Barabino B, Maternini G (2021) Survey on e-powered micro personal mobility vehicles: exploring current issues towards future developments. Sustainability 13(7). https://doi.org/10.3390/su13073692

Ballo L, Meyer de Freitas L, Meister A, Axhausen K W (2023) The e-bike city as a radical shift toward zero-emission transport: sustainable? Equitable? Desirable? J Trans Geogr 111. https://doi.org/10.1016/j.jtrangeo.2023.103663

Ballo L, Raubal M, Axhausen K W (2024) Designing an e-bike city: an automated process for network-wide multimodal road space reallocation. J Cycl Micromobility Res 2. https://doi.org/10.1016/j.jcmr.2024.100048

Bretones A, Marquet O (2022) Sociopsychological factors associated with the adoption and usage of electric micromobility. Lit Rev Trans Policy 127:230–249. https://doi.org/10.1016/j.tranpol.2022.09.008

Cai J, Zhao J, Liu J, Shen K, Li X, Ye Y (2020) Exploring factors affecting the yellow-light running behavior of electric bike riders at urban intersections in China. J Adv Transp. https://doi.org/10.1155/2020/8573232

Cao Z (2025) Understanding how street environment affects e-scooter mode choice through travel experience. Cities 158. https://doi.org/10.1016/j.cities.2024.105511

Cicchino JB, Chaudhary NK, Solomon MG (2023) How are e-scooter speed-limiter settings associated with user behavior? Observed Speeds and Road, Sidewalk, and Bike Lane Use in Austin, TX, and Washington, D.C. Transp Res Record 2678(8):171–183. https://doi.org/10.1177/03611981231214518

Comi A, Polimeni A (2024) Assessing potential sustainability benefits of micromobility: a new data driven approach. Eur Trans Res Rev 16(19). https://doi.org/10.1186/s12544-024-00640-6

Cubells J, Miralles-Guasch C, Marquet O (2023a) Gendered travel behaviour in micromobility? Travel speed and route choice through the lens of intersecting identities. J Trans Geogr 106. https://doi.org/10.1016/j.jtrangeo.2022.103502

Cubells J, Miralles-Guasch C, Marquet O (2023b) E-scooter and bike-share route choice and detours: modelling the influence of built environment and sociodemographic factors. J Trans Geogr 111. https://doi.org/10.1016/j.jtrangeo.2023.103664

Curtis C, Ma X, Reardon L, Legacy C, Stone J (2025) Micro-mobility: a policy vacuum in Australia. Urban Policy Res 43(2):103–120. https://doi.org/10.1080/08111146.2025.2505898

de Kruijf J, Ettema D, Dijst M (2019) A longitudinal evaluation of satisfaction with e-cycling in daily commuting in the Netherlands. Travel Behav Soc 16:192–200. https://doi.org/10.1016/j.tbs.2018.04.003

Ding C, Cao X, Dong M, Zhang Y, Yang J (2019) Non-linear relationships between built environment characteristics and electric-bike ownership in Zhongshan, China. Transp Res Part D Transp Environ 75:286–296. https://doi.org/10.1016/j.trd.2019.09.005

Fan Z, Harper C D (2022) Congestion and environmental impacts of short car trip replacement with micromobility modes. Transp Res Part D Trans Environ 103. https://doi.org/10.1016/j.trd.2022.103173

Fazio M, Giuffrida N, Le Pira M, Inturri G, Ignaccolo M (2021) Planning suitable transport networks for e-scooters to foster micromobility spreading. Sustainability, 13(20). https://doi.org/10.3390/su132011422

Folco P, Gauvin L, Tizzoni M, Szell M (2022) Data-driven micromobility network planning for demand and safety. Environ Plann B Urban Analytics City Sci 50(8):2087–2102. https://doi.org/10.1177/23998083221135611

Fyhri A, Heinen E, Fearnley N, Sundfør HB (2017) A push to cycling—exploring the e-bike's role in overcoming barriers to bicycle use with a survey and an intervention study. Int J Sustain Transp 11(9):681–695. https://doi.org/10.1080/15568318.2017.1302526

Gössling S (2020) Integrating e-scooters in urban transportation: problems, policies, and the prospect of system change. Transp Res Part D Trans Environ 79. https://doi.org/10.1016/j.trd.2020.102230

Hallberg M, Kjær Rasmussen T, Rich J (2021) Modelling the impact of cycle superhighways and electric bicycles. Transp Res Part A Policy Pract 149:397–418. https://doi.org/10.1016/j.tra.2021.04.015

Hossein Sabbaghian M, Llopis-Castelló D, García A (2023) A safe infrastructure for micromobility: the current state of knowledge. Sustainability 15. https://doi.org/10.3390/su151310140

Hu Y, Zhao M, Zhao Z (2024) Uncovering heterogeneous effects of link-level street environment on e-bike and e-scooter usage. Transp Res Part D Trans Environ 136. https://doi.org/10.1016/j.trd.2024.104477

Jin ST, Wang L, Sui D (2023) How the built environment affects E-scooter sharing link flows: a machine learning approach. J Trans Geogr 112. https://doi.org/10.1016/j.jtrangeo.2023.103687

Kazemzadeh K (2025) Assessing e-scooter rider safety perceptions in shared spaces: evidence from a video experiment in Sweden. Accid Anal Prev 211. https://doi.org/10.1016/j.aap.2024.107874

Kazemzadeh K, Ronchi E (2021) From bike to electric bike level-of-service. Transp Rev 42(1):6–31. https://doi.org/10.1080/01441647.2021.1900450

Kazemzadeh K, Sprei F (2022) Towards an electric scooter level of service: a review and framework. Travel Behav Soc 29:149–164. https://doi.org/10.1016/j.tbs.2022.06.005

Khavarian K, Vosough S, Roncoli C (2024) Bike users' route choice behaviour: expectations from electric bikes versus reality in Greater Helsinki. J Cycl Micromobility Res 2. https://doi.org/10.1016/j.jcmr.2024.100045

Kwon N, Chang I, Lee J, Ahn S (2024) Analysis of E-scooter risk factors by road types on different speed levels. KSCE J Civ Eng 28(8):3533–3542. https://doi.org/10.1007/s12205-024-1335-6

Leger SJ, Dean JL, Edge S, Casello JM (2019) "If I had a regular bicycle, I wouldn't be out riding anymore": Perspectives on the potential of e-bikes to support active living and independent mobility among older adults in Waterloo, Canada. Transp Res Part A Policy Pract 123:240–254. https://doi.org/10.1016/j.tra.2018.10.009

Lin X, Wells P, Sovacool BK (2017) Benign mobility? Electric bicycles, sustainable transport consumption behaviour and socio-technical transitions in Nanjing, China. Transp Res Part A Policy Pract 103:223–234. https://doi.org/10.1016/j.tra.2017.06.014

Maurer FL, Meister A, Axhausen KW (2025) Cycling speed profiles from GPS data: Insights for conventional and electrified bicycles in Switzerland. J Cycl Micromobility Res 5. https://doi.org/10.1016/j.jcmr.2025.100077

Meister A, Felder M, Schmid B, Axhausen KW (2023) Route choice modeling for cyclists on urban networks. Transp Res Part A Policy Pract 173. https://doi.org/10.1016/j.tra.2023.103723

Miao L, Liu F, Deng Y (2025) Analysis of traffic conflicts on slow-moving shared paths in Shenzhen, China. Sustainability 17(9). https://doi.org/10.3390/su17094095

Mitropoulos L, Stavropoulou E, Tzouras P, Karolemeas C,Kepaptsoglou K (2023) E-scooter micromobility systems: review of attributes and impacts. Transp Res Interdisc Perspect 21:100888. https://doi.org/10.1016/j.trip.2023.100888

Moher D, Liberati A, Tetziaff J, Altman DG (2009) Preferred reporting items for systematic reviews and meta-analyses: the PRISMA statement. J Clin Epidemiol 62:1006–1012. https://doi.org/10.1371/journal.pmed.1000097

Moinse D (2025) Exploring the relationship between perceived bikeability and gender-inclusive micromobility usage: a study across 53 French cities. Transp Res Part A Policy Pract 193. https://doi.org/10.1016/j.tra.2025.104379

Moran M (2021) Drawing the map: the creation and regulation of geographic constraints on shared bikes and e-scooters in San Francisco, CA. J Trans Land Use 14(1):197–218. https://doi.org/10.5198/jtlu.2021.1816

Nematchoua M, Deuse C, Cools M, Reiter S. (2020) Evaluation of the potential of classic and electric bicycle commuting as an impetus for the transition towards environmentally sustainable cities: a case study of the university campuses in Liege, Belgium. Renew Sustain Energy Rev 119. https://doi.org/10.1016/j.rser.2019.109544

Porojkow I, Lißner S (2024) Friend or foe? A spatial approach to overlay bicycle and scooter trajectories. Eur Trans Res Rev 16(32). https://doi.org/10.1186/s12544-024-00657-x

Rainieri G, Carra M, Richiedei A, Pezzagno M (2024) Evaluating active mobility: enhancing the framework for social sustainability. TeMA J Land Use, Mobility Environ (3):113–128. https://doi.org/10.6093/1970-9870/10912

Rérat P (2021) The rise of the e-bike: towards an extension of the practice of cycling? Mobilities 16(3):423–439. https://doi.org/10.1080/17450101.2021.1897236

Rich J, Jensen AF, Pilegaard N, Hallberg M (2021) Cost-benefit of bicycle infrastructure with e-bikes and cycle superhighways. Case Stud Trans Policy 9(2):608–615. https://doi.org/10.1016/j.cstp.2021.02.015

Romm D, Chávez JAJ, Kinman L, Salsabilian P, McKenzie G, Manaugh K (2025) The cars are going to be alright: examining micromobility infrastructure space allocation and potential improvement scenarios in Montréal. J Cycl Micromobility Res 5. https://doi.org/10.1016/j.jcmr.2025.100071

Sim JW, Cho GH (2025) Safety behaviors and collision risks between pedestrians and micromobilities in multi-user virtual reality environments. Transport Res F Traffic Psychol Behav 112:342–360. https://doi.org/10.1016/j.trf.2025.04.012

Schumann HH, Haitao H, Meister A, Natapov A, Quddus M (2025) City-scale GPS data reveals impact of spatial configuration and dedicated infrastructure on e-scooter route choice. Sci Rep 15. https://doi.org/10.1038/s41598-025-06938-2

Van Cauwenberg J, de Geus B, Deforche B (2018) Cycling for transport among older adults: health benefits, prevalence, determinants, injuries and the potential of e-bikes. In: Curl A, Musselwhite C (eds) Geographies of transport and ageing. Palgrave Macmillan, Cham. https://doi.org/10.1007/978-3-319-76360-6_6

Van den Steen N, de Geus B, Cappelle J, Vanhaverbeke L (2022) Cycling infrastructure for All EPACs included? World Electr Veh J 13(5). https://doi.org/10.3390/wevj13050074

Ventura R, Ghirardi A, Vetturi D, Maternini G, Barabino B (2024) Comparing the vibrational behavior of e-kick scooters and e-bikes: evidence from Italy. Int J Trans Sci Technol 16:68–86. https://doi.org/10.1016/j.ijtst.2023.10.010

van Lierop D, Soemers J, Hoeke L, Liu G, Chen Z, Ettema D, Kruijf J (2020) Wayfinding for cycle highways: Assessing e-bike users' experiences with wayfinding along a cycle highway in the Netherlands. J Trans Geogr 88. https://doi.org/10.1016/j.jtrangeo.2020.102827

Yu Y, Jiang Y, Qiu N, Guo H, Han X, Guo Y (2022) Exploring built environment factors on e-bike travel behavior in urban China: a case study of Jinan. Front Public Health 10. https://doi.org/10.3389/fpubh.2022.1013421

Zhao C, Nielsen TAS, Olafsson AS, Carstensen TA, Meng X (2018) Urban form, demographic and socio-economic correlates of walking, cycling, and e-biking: evidence from eight neighborhoods in Beijing. Transp Policy 64:102–112. https://doi.org/10.1016/j.tranpol.2018.01.018

Zhang Y, Kasraian D, van Wesemael P (2025) Exploring the built environment's impact on e-bikes: longitudinal evidence from the Netherlands. Transp Res Part D Trans Environ 141. https://doi.org/10.1016/j.trd.2025.104643

Zhang Y, Kasraian D, van Wesemael P (2023) Built environment and micro-mobility: a systematic review of international literature. J Trans Land Use 16(1):293–317. https://doi.org/10.5198/jtlu.2023.2266

Zhang Y, Li Y, Yang X, Liu Q, Li C (2013) Built environment and household electric bike ownership: insights from Zhongshan metropolitan area, China. Trans Res Record 2387(1):102–111. https://doi.org/10.3141/2387-12

Zhu C, Susskind J, Chernicoff W, Cong C, Berger AM (2025) Advancing in-neighborhood micromobility networks: Decarbonizing car-centric urbanization through a multimodal network planning framework Cities 161:105858. https://doi.org/10.1016/j.cities.2025.105858

Zuniga-Garcia N, Juri NR, Perrine KA, Machemehl RB (2021) E-scooters in urban infrastructure: understanding sidewalk, bike lane, and roadway usage from trajectory data. Case Stud Trans Policy 9(3):983–994. https://doi.org/10.1016/j.cstp.2021.04.004

Toward a Spatially Equitable Diffusion of Electric Mobility. Empirical Evidence in the Milan Urban Region

Paola Pucci and Luis Leal Pinho De Morais

Abstract E-mobility plays a key role in enabling a post-car city scenario, which aims to reduce car dependency through integrated transport and land-use strategies, alongside the promotion of active mobility, e-shared mobility services, e-transport networks, and related digital platforms. Two critical issues have emerged in advancing the e-mobility transition. First is the need for targeted incentives and policies that support the diffusion of electric vehicles (EVs), grounded in socio-spatial conditions. Second is the planning and deployment of a suitable charging infrastructure network that not only meets the spatial efficiency of supply and demand but also ensures integration with public transport systems, shared mobility services, micromobility, and active travel modes. Focusing on the Milan urban region (Northern Italy), this chapter investigates the spatial conditions influencing EV adoption and introduces a novel indicator—the "power gap"—to assess the adequacy of public charging infrastructure by comparing its supply with the demand derived from EV registrations, within a scenario-based perspective. This approach aims to support policymakers and stakeholders with actionable insights and strategic recommendations to guide the development of a more sustainable, efficient, and equitable charging network, thereby accelerating the broader transition to low-carbon urban mobility.

1 Introduction: E-Mobility and the Post-Car City

E-mobility has been widely regarded for some years as a promising condition for a sustainable mobility transition, particularly if electricity generation shifts from fossil fuels to renewable energy sources (Crabtree 2019; Rietmann et al. 2020; Thiel et al. 2019; Zhang and Fujimori 2020). While electric vehicles (EVs) offer potential

P. Pucci (✉) · L. L. P. De Morais
Department of Architecture and Urban Studies, Politecnico di Milano, Milano, Italy
e-mail: paola.pucci@polimi.it

L. L. P. De Morais
e-mail: luisleal.morais@polimi.it

© The Author(s), under exclusive license to Springer Nature Switzerland AG 2026
R. Moscarelli et al. (eds.), *The Future of Slow Mobility*, Research for Development,
https://doi.org/10.1007/978-3-032-23660-9_5

sustainability advantages, these benefits depend not only on the integration of renewable electricity sources, adequate charging infrastructure, but also on the supportive, integrated transport systems, such as shared mobility and public transport options within a multimodal transport scenario.

This is necessary also because the pathway "mobility with lower emissions" decoupling mobility from emissions thanks to a technology fix (Givoni 2013, p. 209), in the case of EVs, risks not obtaining appreciable results, due to some critical takes on electric mobility, such as the high energy-intensive production required for EV batteries, or traffic congestion, road accidents, and the widespread, parasitic appropriation of public space by the private uses of the (e)cars (Buberger et al 2022).

Hence, e-mobility can play a key role in promoting sustainable urban mobility, only if it is embedded within a broader mobility ecosystem, where the role of the private vehicle changes, along with travel habits. The shift towards electromobility should thus be conceived as part of a wider vision—one that incorporates innovative solutions to enhance already sustainable modes of transport, including public transit, shared mobility, and active travel, along with policies aimed at fostering intramodality and proximity-based accessibility (Moreno et al. 2021; Gower and Grodach, 2022; Allam et al. 2022; Pucci and Vecchio, 2024).

This approach aligns with what has been referred to as a "post-car city" scenario (Beyazit 2013), which aims to reduce car dependency through integrated transport and land-use strategies, alongside the promotion of active mobility, e-shared mobility services, e-transport networks, and related digital platforms (Coppola and Pucci 2023).

Efforts toward a post-car city are increasingly reflected in urban agendas worldwide. These initiatives are driven by the recognition that cities need not only different types of vehicles (e.g., electric, hydrogen-powered, or autonomous), but also fewer vehicles overall (Burns 2013; Legacy et al. 2019).

In these experiences, policies that support electric mobility must deal with two critical issues, addressed by this paper. The first concerns the need for targeted incentives and policy measures to support the adoption of electric vehicles (EVs), tailored to the socio-spatial contexts that may hinder their uptake. The second involves the planning and implementation of a suitable charging infrastructure network—one that not only aligns with the spatial dynamics of supply and demand but is also seamlessly integrated with public transport systems, shared mobility services, micromobility options, and active travel modes. Such integration facilitates intermodal travel, reduces reliance on private car ownership, and supports a more efficient, flexible, and sustainable mobility ecosystem—given the centrality of mobility hubs as places where multimodality can occur.

2 Targeted Incentives and Socio-spatial Conditions for e-Mobility

For a long time, research into the diffusion of electric vehicles (EVs) has primarily focused on scenario construction to assess EV adoption trajectories over the medium term, combining spatial and temporal aspects of EV demand (Higgins et al. 2012; Kihm and Trommer, 2014; Zubaryeva et al. 2012). Several studies have also mapped EV registrations and identified residential locations and potential EV adopters (Campbell et al. 2012; Morton et al. 2018), as well as the optimal locations for public charging infrastructure (Bailey et al. 2015; Namdeo et al. 2014).

By focusing mainly on stated preferences or historical EV registration data, these studies typically find that EV uptake is concentrated in highly urbanised areas where income, education, and home ownership levels are relatively high. However, in such contexts, the one-to-one replacement of internal combustion engine vehicles (ICEVs) with EVs is not a viable long-term solution to issues such as traffic congestion and the excessive use of public space by private cars (Kester et al. 2020; Newman et al. 2014; Plötz et al. 2014).

Studies offering overviews of the benefits and challenges of Nordic electric mobility across urban, suburban, and rural areas provide additional valuable insights for addressing diverse territorial and user profiles. These findings are particularly useful for guiding the EV transition in specific spatial contexts, including rural areas (Kester et al. 2020; Sovacool et al. 2019), highlighting the urgency of investigating the spatial patterns of electric vehicle (EV) demand. This is a key issue for two main reasons: first, to ensure that even peri-urban and car-dependent areas are equipped with adequate recharging infrastructure; and second, to guide the allocation of incentives and subsidies in a more equitable and efficient manner.

Empirical evidence shows that policy measures will be more effective and fairer if they are informed by the sociospatial conditions that give rise to differentiated mobility needs and demands (Pucci 2021). Addressing these sociospatial conditions is also crucial for shaping diversified, context-specific EV policies that can help manage concerns about the environmental sustainability of electric mobility. This is particularly important given the contradictory effects of undifferentiated EV purchase incentives, often justified by a generalized narrative about the positive impacts of EV diffusion. Such a narrative overlooks the need to understand the varied territorial and social contexts that influence how the EV transition unfolds across different regions.

Considering spatial and social conditions for the diffusion of electric vehicles means dealing with at least three differentiated spatial perspectives (Pucci 2021). In high-density urban settlements with strong public transport networks, policy measures should prioritise the reduction of private car ownership. Rather than promoting the one-to-one substitution of internal combustion engine vehicles (ICEVs) with EVs, policies should encourage the shared use of electric vehicles, increase vehicle occupancy rates, and foster multimodal integration. These strategies should be accompanied by car restriction policies, such as extending congestion charges to private EVs. Within this framework, financial incentives should be redirected to support EV-sharing schemes and the development of fast-charging infrastructure, strategically located along major transport corridors, at multimodal hubs, and near key points of interest (POIs). Regulatory measures should be designed to discourage the private use of EVs by limiting benefits such as free parking, access to congestion charge exemptions, or use of bus lanes to shared electric vehicles only. This would align EV policies with broader goals of urban sustainability, equitable access, and reduced car dependency, supporting multimodality among available sustainable transport modes, and giving centrality to mobility hubs. By integrating diverse transport modes at strategic urban locations, mobility hubs facilitate seamless intermodal travel and promote shared mobility solutions, fostering more inclusive and sustainable urban transport systems (Banister 2008). As nodes where physical integration, intelligent technologies, and data-driven systems converge, mobility hubs enable users to switch efficiently between different mobility options (Graf et al. 2022). Their evolving typologies (Weustenenk and Mingardo 2023; Geurs et al. 2023) have gained prominence in urban mobility policies (Hamburg-Nord 2023; Minneapolis Metropolitan Council 2021; Portland Bureau of Transportation 2020), also for supporting a fair and sustainable e-transition, in particular in dense-urbanised contexts (CoMoUK 2019).

In medium-density settlements characterized by infrastructure corridors and transport interchanges—where the "last mile" issue continues to hinder a shift from private to public mobility—policy measures should focus on the development of mobility hubs around suburban railway stations. These hubs would be equipped with dedicated parking facilities and services for electric vehicles, aiming to facilitate modal transfer from private cars to public transport, as well as the integration between e-bikes and public transport services. By integrating EV infrastructure into transport systems, such policies can enhance intermodality, thereby increasing the efficiency and attractiveness of sustainable mobility options in suburban contexts. In addition, these e-shared mobility hubs can provide not only multiple transportation options but also non-transport and public realm components, which are particularly relevant in peri-urban contexts, fostering multimodality (Fanchao and Gonçalo 2020; Graf et al. 2022).

In low-density areas—where the high intensity of car use can maximize the environmental benefits of EV adoption—policy measures should aim at "decarbonizing" mobility by lowering the barriers to EV uptake. This includes providing financial support for e-vehicle purchases and deploying fast-charging infrastructure across the territory. Although monetary incentives are among the most appreciated by potential buyers, and public investment in charging infrastructure has been found to be more effective than direct purchase subsidies (Lieven 2015), the actual impact of these incentives varies significantly depending on spatial conditions, local travel patterns, regulatory frameworks, and consumer preferences (Hardman 2019).

To ensure that the transition to electric mobility is both sustainable and equitable, policies must be sensitive to these socio-spatial contexts to provide multimodal transport solutions, able to integrate e-vehicles into the broad transport system.

3 Planning for a Smart Charging Infrastructure

Ensuring a fair transition to electric mobility requires not only policies and incentives focused on the peculiar socio-spatial conditions, but also well-planned charging infrastructure. Poor infrastructure planning can reduce convenience for users and slow EV adoption. Therefore, considering socio-spatial characteristics in the planning and location of charging stations is key to supporting an equitable and effective shift to electric vehicles. As is known, charging infrastructure is the essential supporting service that enables the development of electric mobility. In cases of poorly planned infrastructure, insufficient charging stations often coincide with reduced convenience for EV users, thereby hindering adoption (Liang et al. 2020). Therefore, studying the primary spatial localization methodologies for electric vehicle charging stations (EVCSs) can serve as a valuable tool in promoting a fair and efficient transition to e-mobility within contemporary urban contexts.

The importance of integrating spatial analysis and infrastructure planning to ensure equitable access and system efficiency in the transition to electric mobility has prompted the use of geographic information systems (GIS) methodologies, traditionally applied to the spatial planning of energy systems, to the planning and deployment of EVCS networks (Ghosh and Musti 2020).

Three main approaches in the literature, characterize the body of research on electric vehicle charging station (EVCS) site selection.

Demand-focused studies concentrate exclusively on estimating future demand, typically based on projections of EV adoption, and on sizing charging networks accordingly. They often adopt a "blank page" approach, planning EVCS networks from scratch while overlooking the existing charging infrastructure. Examples include the works of Sadeghi-Barzani et al. (2014), Namdeo et al. (2014), and Pradhan et al. (2021).

Supply-focused studies describe the structure and characteristics of existing EVCS networks. They often employ functional data analysis and clustering techniques to analyze energy demand based on actual charging transactions. Notable examples include Pallonetto et al. (2020), Lucas et al. (2018), and Straka and Buzna (2019).

Integrated demand and supply studies adopt a more comprehensive perspective, analysing both demand and supply through GIS-based multi-criteria decision analysis (MCDA). They incorporate geographic, economic, and urban criteria into the site selection process. Some studies also consider factors such as charging station capacity and power (Csiszár et al., 2020) or local EV registration data (Fei et al. 2021). Representative examples include Erbaş et al. (2018), Kaya et al. (2020), and Kaya et al. (2021).

Looking at the methodological approaches, a wide variety of methods are employed in the spatial planning of EV charging infrastructure. Geographic overlap methods, such as the one used in the present study, combine multiple heat maps to identify optimal locations through spatial intersections. These methods are computationally efficient and generate outputs that are accessible and easily interpretable by policymakers. In contrast, simulation-based and optimization algorithms—including genetic algorithms and particle swarm optimization—offer more abstract and computationally intensive solutions. While potentially powerful, these methods rely heavily on model assumptions and constraints, often resulting in outputs that are less transparent and harder to interpret. For a comprehensive review of EVCS placement methodologies, see Pagany, Ramirez Camargo et al. (2019). Case studies on EVCS placement span a global context, illustrating the universal nature of this planning challenge. The scale of analysis varies widely: some studies address national-level planning, such as in Ireland and the Netherlands (Pallonetto et al. 2020; Lucas et al. 2018); others focus on multiple urban locations, such as Modena and Florence (De Gennaro et al., 2014). However, the majority of research is conducted at the city or sub-city level, including municipal or district-scale planning (e.g., Sadeghi-Barzani et al. 2014; Rodrigues et al. 2019; Guler and Yomralioglu 2020).

In most of these approaches, a key issue is the discrepancy between administrative boundaries and spatial scale of the analyses. For instance, some county-level studies—such as Tyne and Wear or parts of Ohio —are in fact smaller in area than certain city or district-level studies, like Ankara or the Changping district in Beijing (Maia et al., 2015; Vazifeh et al. 2019).

This research, taking into account the findings from existing literature, proposes an indicator expressed in International System (SI) units—called Charging Gap Indicator. This approach allows for a simpler comparison between demand and supply across different areas of the case study, which focuses on the Milan urban region. Demand is estimated based on electric vehicle consumption, using an empirical dataset of electric vehicle registrations, while the supply analysis relies on a global dataset of electric vehicle charging stations, taking into account the power and capacity of the charging points.

4 The Charging Gap Indicator

4.1 Current State of EV Diffusion and Infrastructure in the Milan Urban Region

A well-functioning electric vehicle (EV) ecosystem depends on the coordinated development of both vehicles and charging infrastructure. Historically, this posed a "chicken-and-egg" dilemma—limited charging stations hindered EV adoption, while investment in infrastructure was deterred by low EV numbers (Coffman et al. 2017). Cities like Milan have addressed this by promoting EV uptake and expanding public charging networks in tandem (Ledna et al. 2022).

The Milan Urban Region (MUR), with its dense, polycentric structure and heavy car dependence, offers a valuable case study for understanding the spatial diffusion of electric mobility in a car-producing South European country. The MUR spans 970 municipalities across Lombardy, eastern Piedmont, and northern Emilia-Romagna. The spatial complexity and governance challenges of the MUR offer a nuanced lens

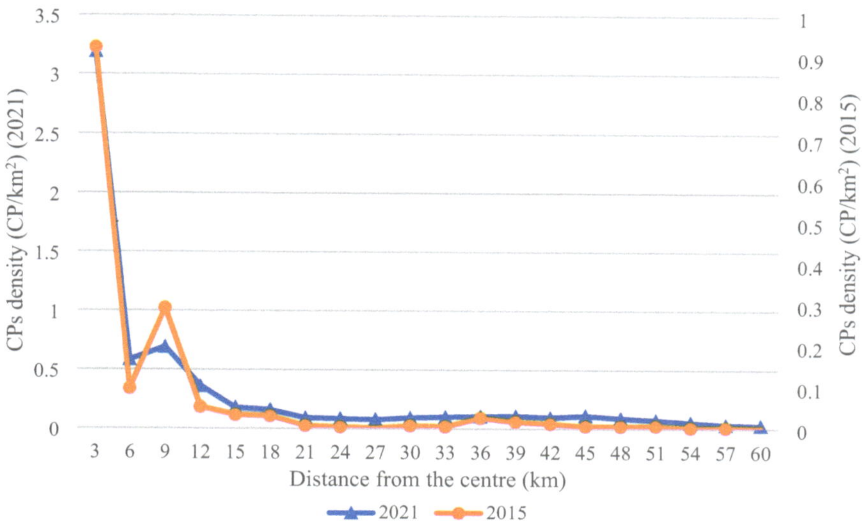

Fig. 1 The CP density gradient in the MUR 2015 and 2021 elaborated from OCM data

to investigate how electric mobility can develop equitably in a region still in transition toward sustainable transport.

Early EV adoption in the Milan Urban Region was concentrated in Milan and Bergamo, gradually spreading outward, with most municipalities still reporting fewer than five EVs by 2021. Charging point distribution followed a similar trajectory—initially limited to Milan in 2015, expanding by 2021 but remaining uneven, particularly in the underserved southern areas.

In Fig. 1, a radial density analysis reveals key spatial disparities. CP density peaks within Milan's 0–3 km core—especially in AREA C—but dips sharply in the surrounding 3–6 km ring. This clustering creates congestion in central areas while limiting access in nearby zones.

To evaluate infrastructure adequacy, the EV-to-charging point (EV/CP) ratio offers a more meaningful metric than absolute charger counts, as it accounts for demand. In the MUR, this ratio rose from 1:4 in 2015 to 1:16 in 2021, reflecting EV growth outpacing infrastructure—mirroring trends across the EU (ACEA 2022).

Despite a better EV/CP ratio than the internal combustion engine vehicle-to-fuel station ratio (1:3000), infrastructure distribution remains problematic. Radial analysis in Fig. 2, highlights undersupply in the 3–6 km and 24–30 km rings (ratios of 1:35 and 1:30), while outer polycentric areas perform better. These disparities risk unequal access, especially for users without home charging options.

However, relying solely on the EV/CP ratio is insufficient, because it overlooks key factors like home charging availability, user behaviour, and charger power levels (Metais et al. 2022. To address these limitations, a more comprehensive metric is proposed: the charging gap. This indicator integrates both supply-side elements (e.g.,

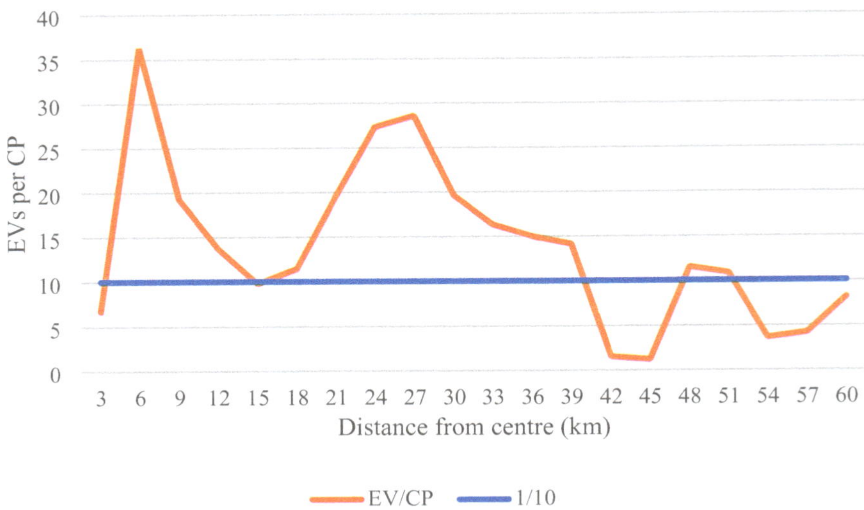

Fig. 2 EV/CP radial density gradient in the MUR in 2021 (source OCM)

charging power, spatial distribution) and demand-side behavior (e.g., trip lengths, usage patterns) to better assess infrastructure adequacy and guide future planning.

4.2 Methodology for Assessing Infrastructure Adequacy

The Public Charging Gap (PCG) is developed to assess the balance between EV charging demand and supply across municipalities in the Milan Urban Region. It is calculated as:

$$\text{PCG} = \text{Charging Supply(CS)} - \text{Charging Demand(CD)}$$

Charging Demand Definition

Charging demand is defined as the daily energy required (in kWh) to charge all EVs registered in a municipality. Demand is modeled based on internal, incoming, and outgoing vehicle flows:

- Internal Flows: EVs live and work/study in the same municipality.
- Outgoing Flows: EVs live in the municipality but commute elsewhere.
- Incoming Flows: EVs commute into the municipality from outside.

Flow-Based Demand Calculation

Each flow type contributes differently to the demand:

- Internal Demand is fully attributed to the home municipality.
- Outgoing Demand is shared between home and work municipalities, proportioned by a day/night charging ratio (R).
- Incoming Demand is similarly split, but the receiving municipality accounts for (1-R) of the load.

Distances between municipalities are approximated using centroid distances (for inter-municipal flows) or diagonal estimates (for intra-municipal flows). EV energy consumption (Wh/km) is adjusted based on average commute speeds using data from the Nissan Leaf 2012.

Total Municipal Demand Formula

The total charging demand D_i for a municipality i is calculated as:

$$D_i = \frac{Ev_i{*}n_{ii}{*}d_{ii}{*}C_i + R\sum_{\substack{j=1\\ j\neq i}}^{M} Ev_i{*}n_{ij}{*}2{*}d_{ij}{*}C_i + (1-R)\sum_{\substack{j=1\\ j\neq i}}^{M} Ev_j{*}n_{ji}{*}2{*}d_{ji}{*}C_j}{t}$$

$$n_{ii} = \frac{N_i}{\sum_{j=1}^{j} N_{ij}}$$

where Ev_i is the number of EVs registered in municipality i (source: Automobile Club Italia), n_{ij} is the percentage of the total cars that are doing that commute. N_i is the absolute number of vehicles travelling inside municipality i, and N_{ij} represents the absolute number of vehicles having a commute starting municipality i. d_{ii} is the average distance travelled inside municipality i. C_i is the electric car consumption (Wh/km), and the driving speed was taken as an influencing variable[1] according to the relationship below that was estimated from different driving cycles of the 2012 Nissan Leaf.

The variables with a double index follow a specific convention: the first position denotes the origin (in this case, home), while the second index signifies the destination (in this case, work). M is the total number of municipalities, and t represents the time needed to charge, where T = 1 h represents a worst-case high-demand scenario.

Public Charging Demand Estimation

Public charging demand (PDi) accounts for the portion of total demand served by publicly accessible infrastructure. This share varies with population density, ranging from 5% (least dense) to 20% (most dense municipalities like Milan). It is normalized by area to yield public charging demand density (NPDi, in kW/km^2).

Public Charging Supply Estimation

Charging supply (CS) is the total available power for EV charging in an area and is calculated as:

[1] Given the reduced scale of the territory under consideration, ambient conditions such as temperature, precipitation, wind, and others were presumed to remain constant throughout.

$$CS = \frac{(\text{Number of Stations} \times \text{Number of Plugs} \times \text{Power per Plug})}{\text{Municipal Area}(\text{km}^2)}$$

Plug counts vary significantly: most stations in the MUR have 1–2 plugs, with an average of 2.2. Power capacities range from 3–150 kW and the average power is 40 kW. In 2020, MUR had 1108 plugs and the total capacity was 73 MW.

The charging gap was calculated using the balanced day/night demand scenario. Areas with negative values (more demand than supply) were marked in red; positive values (more supply than demand) in green; and balanced areas in brown (Fig. 3).

Overall, in terms of the charging gap, Milan and Bergamo, demonstrate adequate supply despite high demand. In contrast, the Southern regions exhibit widespread negative gaps, with limited or no nearby charging alternatives underscoring the urgent need for infrastructure development. Some municipalities showing high positive gaps likely reflect future-proofing efforts. These results should be interpreted with caution due to several limitations in the model. The analysis relies on mobility data from 2011, additionally, the model assumes simplified such as EVs only charging either at home or work and applies uniform energy consumption per municipality.

If the current framework offers a solid foundation for planning future charging infrastructure, with the expected rise in EV adoption, these patterns are likely to shift, and scenario-based approaches will be essential, as will be displayed in the next section.

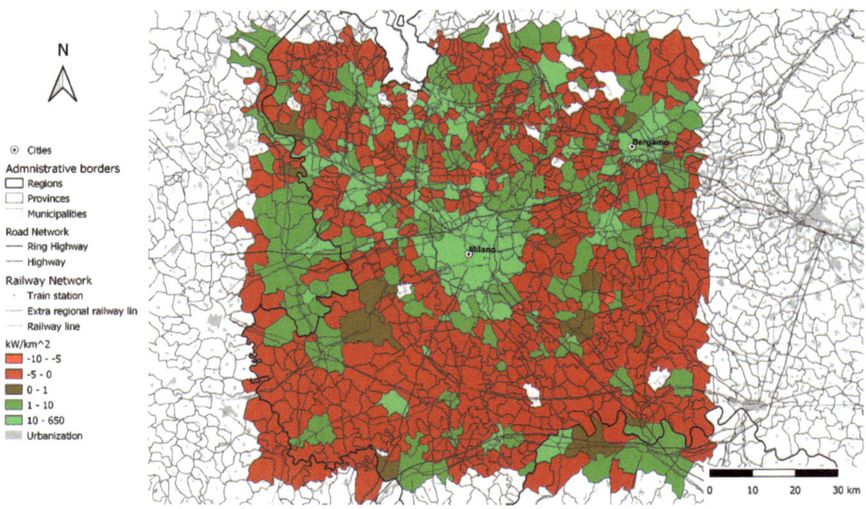

Fig. 3 Spatial distribution of charging gap (kW/km^2) over the MUR territory

5 Findings and Scenario-Based Insights

This section evaluates four key future scenarios by comparing projected 2030 demand with 2020 infrastructure levels. It estimates the resulting charging infrastructure deficit and the investment required to close this gap. Two projections for EV adoption are considered: one based on data from the PNIEC (Piano Nazionale Integrato per l'Energia e il Clima) and MOTUS-E (the Italian association for electric mobility), and another derived by adapting the Dutch EV adoption rate to the Italian context.

In the Business-as-Usual (BaU) Scenario, charging infrastructure evolves based on profitability, and government EV incentives are applied uniformly across the region. Unsurprisingly, this leads to an increase in infrastructure disparities. Milan becomes the most critical municipality. The North MUR area is particularly strained, while the South suffers from large stretches of uncovered areas, restricting access within a wider driving radius. To meet the projected 2030 demand, a 150% increase in public chargers is needed. This expansion raises concerns due to the dense urban fabric of Milan, where clustering chargers may strain both space and the power grid.

In the Eco-Driven (EcD) Scenario, EV incentives are tied to settlement density, encouraging more strategic and equitable adoption, supporting the creation of e-mobility hubs. While Milan remains critical, its charging gap receives a 40% improvement over BaU. The northern critical zones are also less severe. This scenario requires a 125% infrastructure increase. Although this involves a lower investment, it also assumes a smaller EV fleet, trading off total electrification for improved spatial equity.

In the Demand-Driven (DD) Scenario, municipalities guarantee charger placement upon driver request, reducing infrastructure barriers. The result: a 50% larger EV fleet than BaU, leading to wider infrastructure pressure—a 5% increase in overall negative gap area. Milan is again the most critical. To meet demand, a 175% increase is needed. While this implies higher absolute investment, it responds to real demand patterns more inclusively.

A hybrid of EcD and DD, the Accelerated Eco (AcE) Scenario, merges targeted incentives and infrastructure assurance. the North MUR, especially Milan and Brianza, remains a hotspot. However, Bergamo and TEEM/A7-A8 corridor municipalities show resilience with positive charging gaps. To close the gap, a 150% infrastructure expansion is required. This scenario balances the trade-offs of electrification intensity and spatial equity.

This comparative and spatially grounded scenario analysis reveals key tensions between EV growth, charging accessibility, and infrastructure investment. It underlines that EV quantity, more than distribution, drives infrastructure pressure. Milan requires special attention due to its grid and spatial constraints; settlement-based incentives can lower costs and improve system resilience, also organising the supply in well-equipped mobility hubs. In this way, careful planning is needed to avoid reinforcing private car dependence, particularly in well-connected urban cores, where intermodality must play a pivotal role in supporting shared e-mobility solutions.

6 Conclusion. Towards a Sustainable and Inclusive E-Mobility Future

While electric vehicles (EVs) powered by renewable energy can reduce emissions and support sustainability goals, they are a costly solution and offer limited benefits within the current car-dependent systems. Specifically, EVs do not address key externalities such as traffic congestion and road accidents. In contrast, public transport and active modes like walking and cycling remain the most sustainable options. They typically entail lower infrastructure costs and deliver broader social and environmental benefits. Therefore, shifting focus from private vehicle electrification to electrifying public transport and shared transport options is essential for a more effective and equitable transport transition.

In this framework, the planning of mobility hubs capable of supporting intermodality plays a relevant role both in densely urbanized contexts and in peri-urban areas, where last-mile issues affect mobility practices.

The proposed Charging Gap Indicator, while not accounting for the broader negative externalities of electric vehicles, serves as a useful tool for comparing the supply and demand of EV charging infrastructure and addressing mobility policy through an effective distribution able to support intermodality. By employing standardized SI units, it enables consistent, like-for-like comparisons across different geographic regions. The municipal boundaries of Milan were initially identified as having a positive charging gap; however, this condition was reversed in the projected scenarios as the municipality was the most critical across all scenarios.

The eco-driven scenario illustrates that directing EV incentives toward lower-density settlements can improve policy effectiveness. This approach primarily targets high-mileage vehicles while reducing the overall incentive budget, as these areas are less populated and therefore require fewer subsidies. Additionally, in these settlements, public–private partnerships in charging infrastructure can enhance the attractiveness of otherwise less profitable locations for private operators.

In dense urban areas, electric vehicles (EVs) should be implemented primarily as part of shared mobility services in multimodal mobility hubs. This approach can reduce parking demand, enhance integration with public transport, (e)shared mobility options and alleviate the burden of charging infrastructure on individual users.

In the case of Italy, national policies on electric mobility require reform to better support a sustainable transition. Notably, EV incentive schemes should be revised to prioritize battery electric vehicles (BEVs) over plug-in hybrid electric vehicles (PHEVs), which offer more limited environmental benefits. Additionally, investment in digital infrastructure is essential to align public charging availability with actual mobility patterns. A strategic approach to charging deployment should be adopted—promoting fast charging during the day in high-inflow areas (e.g., employment and commercial centers) and supporting overnight charging infrastructure in residential, outward-commuting zones. In addition, rethinking the diffusion of electric vehicles through shared and integrated modes underscores the centrality of combining the e-charging locations strategies with the design of mobility hubs where intermodality can ensure that EV rollout is equitable, efficient, and environmentally effective.

References

ACEA (2022) Enabling factors for alternatively-powered cars and vans in the European Union. https://www.acea.auto/files/Research-Whitepaper-A-European-EV-Charging-Infrastructure-Masterplan.pdf

Allam Z, Bibri SE, Chabaud D, Moreno C (2022) The '15-minute city' concept can shape a net-zero urban future. Humanit Soc Sci Commun 9(1):1–5. https://doi.org/10.1057/s41599-022-01145-0

Bailey J, Miele A, Axsen J (2015) Is awareness of public charging associated with consumer interest in plug-in electric vehicles? Transp Res Part D 36:1–9

Banister D (2008) The sustainable mobility paradigm. Transp Policy 15(2):73–80. https://doi.org/10.1016/j.tranpol.2007.10.005

Beyazit E (2013) Mobility cultures. In: Givoni M, Banister D (eds) Moving towards low carbon mobility. Edward Elgar, Cheltenham, pp 15–26

Bezirk Hamburg-Nord (2023) Carsharing-Ausbau im Bezirk Umsetzung 32 neuer hvv switch Punkte in Hamburg-Nord. Hamburg.De. https://www.hamburg.de/hamburg-nord/pressemitteilungen/17843834/carsharing-ausbau-hvv-switch/

Buberger J, Kersten A, Kuder M, Eckerle R, Weyh T, Thiringer T (2022) Total CO_2-equivalent life-cycle emissions from commercially available passenger cars. Renew Sustain Energy Rev 159:112158. https://doi.org/10.1016/j.rser.2022.112158

Burns LD (2013) Sustainable mobility: a vision of our transport future. Nature 497(7448):181

Campbell AR, Ryley T, Thring R (2012) Identifying the early adopters of alternative fuel vehicles: a case study of Birmingham, United Kingdom. Transp Res Part A Policy Pract 46(8):1318–1327. https://doi.org/10.1016/j.tra.2012.05.004

Coffman M, Bernstein P, Wee S (2017) Electric vehicles revisited: a review of factors that affect adoption. Transp Rev 37(1):79–93. https://doi.org/10.1080/01441647.2016.1217282

CoMoUK. (2019) Mobility Hubs Guidance. www.como.org.uk

Coppola PL, Pucci P (2023) Post car city un modello in cerca di politiche. In: Coppola PL, Pucci P, Pirlo G (eds) Mobilità & Città: verso una post-car cit. Ottavo Rapporto Urban@it sulle città. Il Mulino, Bologna

Crabtree G (2019) The coming electric vehicle transformation. Science 366(6464):422–424

Csiszár C, Csonka B, Földes D, Wirth E, Lovas T (2020) Location optimisation method for fast-charging stations along national roads. J Transp Geogr 88:102833. https://doi.org/10.1016/j.jtrangeo.2020.102833

De Gennaro M, Paffumi E, Scholz H, Martini G (2014) GIS-driven analysis of e-mobility in urban areas: an evaluation of the impact on the electric energy grid. Appl Energy 124:94–116. https://doi.org/10.1016/j.apenergy.2014.03.003

Erbaş M, Kabak M, Özceylan E, Çetinkaya C (2018) Optimal siting of electric vehicle charging stations: a GIS-based fuzzy multi-criteria decision analysis. Energy 163:1017–1031. https://doi.org/10.1016/j.energy.2018.08.140

Fei Y, Kang W, Li L, Wang Z (2021) Why do consumers choose to buy electric vehicles? A paired data analysis of purchase intention configurations. Transp Res Part A Policy Pract 147. https://doi.org/10.1016/j.tra.2021.02.014

Fanchao L, Gonçalo C (2020) Smart shared green mobility hubs. Deliverable 1.1 State-of-the-art related to eHUBS.

Geurs K, Grigolon A, Münze K, Gkiotsalitis K, Duran-Rodas D, Büttner B, Kirchberger C, Pappers J, Martinez Ramirez L, Graf A, Hansel J, Gkrava R, Klementschitz R (2023) The Smarthubs

integration ladder: a conceptual model for the categorisation of shared mobility hubs. Trans Rev 1–28. https://doi.org/10.1080/01441647.2023.2239499
Ghosh K, Musti KSS (2020) Integration of SLAM with GIS to model sustainable urban transportation system: a smart city perspective. In: 12th International conference on computational intelligence and communication networks (CICN), Bhimtal, India, pp 261–267. https://doi.org/10.1109/CICN49253.2020.9242571
Givoni M (2013) Alternative pathways to low carbon mobility. In: Givoni M, Banister D (eds) Moving towards low carbon mobility. Edward Elgar Publishing, Cheltenham
Gower A, Grodach C (2022) Planning innovation or city branding? Exploring how cities operationalise the 20-minute neighbourhood concept. Urban Policy Res 40(1):36–52
Graf A, Hansel J, Kirchberger C, Wien TU, Klementschitz R, Gkiotsalitis K (2022) A multidimensional mobility hub typology and inventory. SmartHubs Deliverable D 2.1.
Guler D, Yomralioglu T (2020) Suitable location selection for the electric vehicle fast charging station with AHP and fuzzy AHP methods using GIS. Ann GIS 26(2):169–189. https://doi.org/10.1080/19475683.2020.1737226
Hardman S (2019) Understanding the impact of reoccurring and non-financial incentives on plug-in electric vehicle adoption—a review. Transp Res Part A Policy Pract 119:1–14
Higgins A, Paevere P, Gardner J, Quezada G (2012) Combining choice modelling and multi-criteria analysis for technology diffusion: an application to the uptake of electric vehicles. Technol Forecast Soc Chang 79(8):1399–1412. https://doi.org/10.1016/j.techfore.2012.04.008
Kaya Ö, Alemdar KD, Campisi T, Tortum A, Çodur MK (2021) The development of decarbonisation strategies: a three-step methodology for the suitable analysis of current EVCS locations applied to Istanbul, Turkey. Energies 14(10). https://doi.org/10.3390/en14102756
Kaya Ö, Tortum A, Alemdar KD, Çodur MY (2020) Site selection for EVCS in Istanbul by GIS and multi-criteria decision-making. Transp Res Part D Transp Environ 80:102271. https://doi.org/10.1016/j.trd.2020.102271
Kester J, Sovacool BK, Noel L, Zarazua de Rubens G (2020) Rethinking the spatiality of Nordic electric vehicles and their popularity in urban environments: Moving beyond the city? J Transp Geogr 82:102610. https://doi.org/10.1016/j.jtrangeo.2020.102610
Kihm A, Trommer S (2014) The new car market for electric vehicles and the potential for fuel substitution. Energy Policy 73:147–157. https://doi.org/10.1016/j.enpol.2014.05.021
Ledna C, Muratori M, Brooker A, Wood E, Greene D (2022) How to support EV adoption: tradeoffs between charging infrastructure investments and vehicle subsidies in California. Energy Policy 165:112931. https://doi.org/10.1016/j.enpol.2022.112931
Legacy C, Ashmore D, Scheurer J, Stone J, Curtis C (2019) Planning the driverless city. Transp Rev 39(1):84–102
Liang Y, Guo C, Yang J, Ding Z (2020) Optimal planning of charging station based on discrete distribution of charging demand. IET Gener Transm Distrib 14(6):965–974. https://doi.org/10.1049/iet-gtd.2019.0292
Lieven T (2015) Policy measures to promote electric mobility—a global perspective. Transp Res Part A Policy Pract 82:78–93. https://doi.org/10.1016/j.tra.2015.09.008
Lucas A, Prettico G, Flammini MG, Kotsakis E, Fulli G, Masera M (2018) Indicator-based methodology for assessing EV charging infrastructure using exploratory data analysis. Energies 11(7). https://doi.org/10.3390/en11071869
Maia SM, Teicher H, Meyboom A (2015) Infrastructure as social catalyst: electric vehicle station planning and deployment. Technol Forecast Soc Chang. https://doi.org/10.1016/j.techfore.2015.09.020
Metais MO, Jouini O, Perez Y, Berrada J, Suomalainen E (2022) Too much or not enough? Planning electric vehicle charging infrastructure: A review of modeling options. Renew Sustain Energy Rev 153:111719. https://doi.org/10.1016/j.rser.2021.111719
Minneapolis Metropolitan Council (2021) Mobility hub survey analysis results-high-level findings. https://metrocouncil.org/Transportation/Publications-And-Resources/Planning/MISCELLANEOUS-DOCUMENTS/2021-Mobility-Hub-Survey-Analysis-Results.aspx

Moreno C, Allam Z, Chabaud D, Gall C, Pratlong F (2021) Introducing the 15-minute city: sustainability, resilience and place identity in future post-pandemic cities. Smart Cities 4(1):93–111. https://doi.org/10.3390/smartcities4010006

Morton C, Anable J, Yeboah G, Cottrill C (2018) The spatial pattern of demand in the early market for electric vehicles: evidence from the United Kingdom. J Transp Geogr 72:119–130. https://doi.org/10.1016/j.jtrangeo.2018.08.010

Namdeo A, Tiwary A, Dziurla R (2014) Spatial planning of public charging points using multi-dimensional analysis of early adopters of electric vehicles for a city region. Technol Forecast Soc Chang 89:188–200. https://doi.org/10.1016/j.techfore.2013.08.032

Newman D, Wells P, Donovan C, Nieuwenhuis P, Davies H (2014) Urban, sub-urban or rural: where is the best place for electric vehicles? Int J Automot Technol Manag 14(3–4):306–323. https://doi.org/10.1504/IJATM.2014.065295

Pagany R, Ramirez Camargo L, Dorner W (2019) A review of spatial localization methodologies for the electric vehicle charging infrastructure. Int J Sustain Transp 13(6):433–449. https://doi.org/10.1080/15568318.2018.1481243

Pallonetto F, Galvani M, Torti A, Vantini S (2020) A framework for analysis and expansion of public charging infrastructure under fast penetration of electric vehicles. World Electr Veh J 11(1). https://doi.org/10.3390/wevj11010018

Plotz P, Schneider U, Globisch J, Dütschke E (2014) Who will buy electric vehicles? Identifying early adopters in Germany. Transp Res Part A Policy Pract 67:96–109. https://doi.org/10.1016/j.tra.2014.06.006

Portland Bureau of Transportation (2020) Mobility hub typology study. Portland (US)

Pradhan S, Ghose D, Shabbiruddin (2021) Planning and design of suitable sites for electric vehicle charging station: A case study. Int J Sustain Eng 14(3):404–418. https://doi.org/10.1080/19397038.2020.1862347

Pucci P (2021) Spatial dimensions of electric mobility. Scenarios for efficient and fair diffusion of electric vehicles in the Milan Urban Region. Cities 110:103069. https://doi.org/10.1016/j.cities.2020.103069

Pucci P, Vecchio G (eds) (2024) Questioning proximity. Springer, Opportunities and challenges for urban planning and mobility policies. https://doi.org/10.1007/978-3-031-66071-9

Rietmann N, Hügler B, Lieven T (2020) Forecasting the trajectory of electric vehicle sales and the consequences for worldwide CO_2 emissions. J Clean Prod 261:121038

Rodrigues JL, Bolognesi HM, Melo JD, Heymann F, Soares FJ (2019) Spatiotemporal model for estimating electric vehicles adopters. Energy 183(C):788–802. https://ideas.repec.org/a/eee/energy/v183y2019icp788-802.html

Sadeghi-Barzani P, Rajabi-Ghahnavieh A, Kazemi-Karegar H (2014) Optimal fast charging station placing and sizing. Appl Energy 125(C):289–299. https://EconPapers.repec.org/RePEc:eee:appene:v:125:y:2014:i:c:p:289-299

Sovacool BK, Kester J, Noel L, Zarazua de Rubens G (2019) Income, political affiliation, urbanism and geography in stated preferences for electric vehicles (EVs) and vehicle-to-grid (V2G) technologies in Northern Europe. J Transp Geogr 78:214–229. https://doi.org/10.1016/j.jtrangeo.2019.06.006

Straka M, Buzna Ľ (2019) Clustering algorithms applied to usage related segments of electric vehicle charging stations. Transp Res Procedia 40:1576–1582. https://doi.org/10.1016/j.trpro.2019.07.218

Thiel C, Julea A, Acosta Iborra B, De Miguel EN, Peduzzi E, Pisoni E, Pisoni E, Vilchez JJG, Krause J (2019) Assessing the impacts of electric vehicle recharging infrastructure deployment efforts in the European Union. Energies 12(12):2409

Vazifeh MM, Zhang H, Santi P, Ratti C (2019) Optimizing the deployment of electric vehicle charging stations using pervasive mobility data. Transp Res Part A Policy Pract 121:75–91. https://doi.org/10.1016/j.tra.2019.01.002

Weustenenk AG, Mingardo G (2023) Towards a typology of mobility hubs. J Transp Geogr 106:103514. https://doi.org/10.1016/j.jtrangeo.2022.103514

Zhang R, Fujimori S (2020) The role of transport electrification in global climate change mitigation scenarios. Environ Res Lett 15(3)
Zubaryeva A, Thiel C, Barbone E, Mercier A (2012) Assessing factors for the identification of potential lead markets for electrified vehicles in Europe: expert opinion elicitation. Technol Forecast Soc Chang 79:1622–1637

E-Bike in Extra-Urban Context: A New Era for Slow Tourism

Luca Pirolo and Paolo Pileri

Abstract This chapter analyses the integration between slow tourism and electric cycling, placing it within the broader debate on sustainable tourism and the reduction of environmental impacts linked to tourist mobility. In the face of the climate crisis and the critical issues of mass tourism, slow tourism is emerging as an alternative paradigm, based on low-emission travel practices, the improvement of local contexts, and a more conscious experience of the territories. In this context, cycling tourism is one of the most significant expressions of slow tourism, with significant economic and social benefits, particularly in rural and marginal areas. In recent years, the spread of electric bicycles has deeply changed the sector, expanding the user base and redefining how people access territories. The chapter addresses this process through the conceptual lenses of inclusion and empowerment, distinguishing between the ability to engage a plurality of users and that of making traditionally difficult-to-reach territories accessible. It highlights both the inclusive potential of e-bikes in cycling tourism and the emerging critical issues related to access costs, environmental impacts, safety, and infrastructure dependence. It is therefore argued that electric cycling tourism can contribute to the objectives of slow tourism only if it is part of a conscious planning and policy vision capable of combining technological innovation, social equity, and protection of the local areas.

1 Integrating Slow Tourism and E-Mobility

The tourism sector is currently facing an urgent challenge linked to the climate crisis, which requires a deep reflection on development models and mobility practices associated with travel (Peeters et al. 2024; Kaján and Saarinen 2013; Burns and Bibbings 2009; Scott 2011). Growing environmental and ecological concerns are

L. Pirolo (✉)
Department of Electronics, Information, and Bioengineering, Politecnico di Milano, Milan, Italy
e-mail: luca.pirolo@polimi.it

P. Pileri
Department of Architecture and Urban Studies, Politecnico di Milano, Milan, Italy
e-mail: paolo.pileri@polimi.it

© The Author(s), under exclusive license to Springer Nature Switzerland AG 2026
R. Moscarelli et al. (eds.), *The Future of Slow Mobility*, Research for Development,
https://doi.org/10.1007/978-3-032-23660-9_6

closely linked to rising carbon emissions, one of the main causes of global warming. Although these emissions come from various economic sectors, tourism is one of the most significant contributors, particularly due to the spread of traditional mass tourism and energy-intensive mobility patterns (Ceron and Dubois 2003; Gössling 2002; Simpson et al. 2008; Sun et al. 2024). In this context, the promotion of low-emission practices is a central issue for the future of sustainable tourism (Gössling 2000).

One of the emerging responses to these critical issues is slow tourism, a paradigm that proposes a rethinking of the timing, methods, and purposes of travel, emphasising the reduction of environmental impacts, the enhancement of local contexts, a more conscious experience of destinations, and a different experience of perception of places (Dickinson and Lumsdon 2010; Conway and Timms 2010; Matos 2012; Riga 2025). Slow tourism is often presented as an alternative to mass tourism, capable of promoting more sustainable forms of mobility and strengthening the bond between visitors, territories, and local communities (Blanco 2011; Heitmann 2011; Calzati and De Salvo 2017; Pileri and Moscarelli 2021).

Within this framework, cycling tourism is one of the most significant expressions of slow tourism. It is characterised by the use of bicycles as the main means of transport and by a mode of travel in which the route takes on a central importance compared to the final destination. According to a widely accepted definition, cycling tourism is a leisure cycling activity lasting at least 24 h, involving a minimum distance of 150 km and at least one overnight stay away from home (Ritchie 1998). This practice allows travellers to explore different areas, appreciate their natural and cultural heritage, and support local economies by using local services and products. In recent years, cycling tourism has been recognised as a rapidly growing sector, with an increasing number of travellers and a significant economic impact, particularly in rural and peripheral areas (Ciascai et al. 2022; Cecere et al. 2021; Han et al. 2017a; Petroman et al. 2021). Recognized as a rapidly expanding sector, cycling tourism attracts an increasing number of travelers every year (Ciascai et al. 2022; Cecere et al. 2021; Han et al. 2017a) and has a significant economic impact (Petroman et al. 2021; ADFC Bicycle Travel Analysis 2023). *"Slow travellers" contribute to the local economy by stimulating the development of infrastructure (e.g. roads, bicycle rental/ repair shops, and accommodation). At the same time, the development of the service sector (e.g. food shops) [...] offers commercial and employment opportunities* (Han et al. 2017b).

Recently, the cycling tourism sector has undergone further evolution, thanks to the widespread adoption of pedal-assisted bicycles, commonly referred to as e-bikes. These bicycles have transitioned from a niche use to a mainstream practice, enlarging the type of users by age, physical performance, and personal interests. This was also demonstrated by both global market data and the growth of electric cycling tourism in various national contexts (Mitterwallner et al. 2021). Looking to the global electric bicycle market, the sales of electric bikes experienced a 6.2% increase from 2020 to 2025, with forecasts predicting a further growth rate of 3.56% by 2030 (Mordor Intelligence 2020). This transformation potentially expands the user base and changes the way people access areas, significantly affecting the travel experience.

The introduction of electric propulsion technologies represents a significant shift in the way cycling tourism is practised, constituting a real socio-technical transition rather than a simple incremental improvement (Corti et al. 2024; Baltazar et al. 2022). For many years, mass tourism was largely confined to specific places such as historic cities, coastal areas, or alpine landscapes. However, the integration of cycling tourism and electric mobility risks exporting mass tourism to areas where it has never existed before, particularly in environmentally sensitive regions.

Despite growing academic interest in cycling tourism and electric mobility, existing literature tends to focus primarily on individual aspects, such as user benefits, the technological performance of e-bikes, or the economic impact of the sector. However, there are still limited integrated studies on the role of e-bikes in cycling tourism within the paradigm of slow tourism, systematically analysing their implications in terms of social inclusion and territorial empowerment, as well as the critical issues and potential environmental and infrastructural conflicts that arise from them.

Starting from this gap, this paper aims to analyse the role of electric bicycle mobility in cycling tourism through the conceptual lenses of inclusion and empowerment. Inclusion is understood here, in a broad sense, as the ability of tourism practices to involve a variety of users with different physical, social, and experiential characteristics. Empowerment, on the other hand, refers to the capacity to enhance the agency, resilience, and development potential of territories and destinations that are traditionally difficult to access due to their morphological, infrastructural, or locational characteristics. Although closely interrelated, these two concepts refer to distinct analytical dimensions, focusing on users and places, respectively, that are particularly relevant in the context of slow tourism and cycling tourism.

Section 2 analyses how cycling tourism can be an inclusive practice, further expanded by the introduction of electric cycling. Section 3 is dedicated to the theme of territorial empowerment, examining how cycling tourism and e-biking contribute to making traditionally marginal areas accessible. Finally, Sect. 4 explores the critical issues and controversies associated with the integration of cycling tourism and electric cycling.

2 Slow Tourism as an Inclusive Practice

To focus on the concept of slow tourism as an inclusive practice, it is necessary to take a step back and fully grasp the core meaning of slowness. Slowness should not be interpreted as a mere stylistic or promotional device, but rather as a relational and cognitive approach to travelling. It represents a lens through which places are not simply crossed or consumed, but read, interpreted, and gradually understood. In Italian, the word "*lente*" (lens) is embedded in the term "*lentezza*" (slowness), suggesting a semantic and conceptual link between observation and temporality (Pileri 2020). Through slowness, otherwise overlooked details become visible, allowing travellers to engage with the material, historical, and narrative dimensions of the territories they cross, without the burden of a didactic or prescriptive experience

(Christin 2019). Slowness thus enables a form of learning that is light, experiential, and embodied, rather than instructional. This approach encourages a gradual discovery of places, made possible by long presence, attentiveness to the surroundings, and openness to encounter. Travelling slowly means moving through territories without indifference: not passing through to consume, but to absorb. In this sense, slow tourism fosters an immersive and inclusive relationship with places, valuing landscapes, everyday practices, and human interactions as integral components of the travel experience. From this perspective, slowness is not an end in itself, nor a marketing-driven promise of "immersion", but a design principle for tourism practices and infrastructures aimed at enabling meaningful access to places and experiences. As such, slowness can be understood as a potentially inclusive practice, accessible to a wide range of users and capable of reducing experiential inequalities, in contrast to the selective and exclusionary logic often associated with speed.

In the context of contemporary transport systems, access to speed is closely linked to economic availability: as speed increases, so do costs, as is the case with transport, data transmission, or access to leisure services. Speed, therefore, takes on a more selective and exclusive character, while slowness is, at least in part, more accessible. Cycling tourism also requires an initial investment, but this is generally modest and cannot be compared to the costs of accessing faster forms of mobility. This does not imply a demonisation of speed, whose technological and social benefits are widely recognised, but rather a claim to the right to choose slow mobility practices that are accessible to a large section of the population. Differentiated access to speed is a factor of social division, which tends to segment society on the basis of income. On the contrary, slowness can be practised by a wider range of people, including men and women, the elderly, families, and, potentially, people with permanent or temporary disabilities, provided that routes are designed according to criteria of inclusiveness. In particular, for cycling routes to be accessible to a variety of cycling modes, such as handbikes, tricycles, tandems, or bicycle rickshaws, the road surface must be smooth and uniform, and the slopes must be moderate. Cycling tourism does not rely on specialised technical skills or competitive training, as it is primarily oriented towards leisure and experience; when supported by appropriate infrastructure, it can be accessible to a wider range of users than is commonly perceived.

In addition to the individual dimension of the tourist experience, the link between cycling tourism and inclusion can also be seen on a regional and community level. A cycling tourism project, if conceived as an integrated system of services, can in fact have a high potential for inclusion. To ensure the sustainability of the cycling tourism ecosystem, understood as its ability to be supported and recognised by the communities it passes through, a wide range of services is required: hospitality, and linked services such as museums, guides, theatres, ecological experiences, maintenance of the territory and infrastructure, technical assistance for bicycles, as well as the promotion of local products. These activities can be designed with the active involvement of the local population, including those who experience temporary or structural forms of social exclusion. In this sense, cycling tourism can help to generate opportunities for economic and social participation for groups that are often marginalised in the labour market, such as former prisoners or people with disabilities. If designed with a clear focus on social inclusion, cycling tourism projects can serve not only as leisure infrastructure, but also as a means of promoting territorial cohesion, particularly when they support stable and meaningful employment opportunities and foster balanced, respectful relationships between tourists and residents. However, such outcomes are not automatic and depend on deliberate and context-sensitive planning choices.

When reflecting on slow tourism, slowness is often associated with the concept of speed limits, in opposition to a society characterised by the myth of acceleration (Rosa 2015), summarised by the expression "time is money". However, slowness can be interpreted as a liberating practice, with its own specific speed. In his work *Energy and Equity* (1974), Ivan Illich identified the threshold of 25 kms per hour as a critical point beyond which equity decreases and the scarcity of time and space increases, due to the progressive monopolisation of public space by motorised transport. Speed thus becomes a new social discriminant, capable of separating those who can afford to "buy" speed from those who are excluded from it. A similar reflection also emerges in travel literature. In his account of his journey along the Po River, Ermanno Rea identifies a speed of around 20 kms per hour as the ideal condition for the tourist experience, capable of maintaining a balance between the material perception and symbolic dimension of the landscape (Rea 2017). Both perspectives converge in identifying 20–25 kms per hour as a threshold within which it is possible to preserve a sensory and contemplative relationship with the environment being traversed.

At these speeds, movement can only be sustained by the body's energy. Cycling tourism is therefore an intrinsically inclusive and energy-free activity. In its traditional form, it does not require external energy supplies, except for stops for rest or food. The bicycle is also the most energy-efficient means of transport compared to walking, under the same external conditions (Illich 1974), and is one of the most equitable modes of transport in terms of energy consumption and affordability.

The introduction of electric cycling into cycling tourism further changes the relationship between slowness and inclusion. Thanks to their pedal assistance, e-bikes help to reduce the physical barriers associated with slopes, distances, and the morphological conditions of the terrain, thereby expanding the pool of potential users (Aprigliano et al. 2025; Weinert et al. 2007). This particularly benefits less trained cyclists, older people, families with children, and occasional users. In this sense, electric cycling can strengthen the inclusive nature of cycling tourism, allowing people with different physical abilities to participate in slow tourism experiences. The expansion of the types of users involved in cycling tourism has significant implications for how territories are crossed and enjoyed, a topic that will be explored in more detail in the next section. At the same time, the introduction of an external energy source entails a change in the relationship between user, vehicle, and infrastructure, raising issues related to equity, energy dependence, and the need for support services.

While traditional cycling tourists are essentially independent of external energy sources, the use of e-bikes introduces a new form of infrastructural dependency linked to the availability of energy for recharging and battery disposal. However, the possibility of using pedal-assisted bicycles even in the absence of electricity, once the battery is depleted, opens up further thoughts on the meaning of autonomy, equity, inclusion and road safety within slow tourism. While these issues do not call into question the inclusive potential of e-bikes, they highlight the need for broader reflection on their role in contemporary cycling tourism, which will be developed in the following sections.

3 Slow Tourism as an Empowering Practice

Cycling tourism is not only an activity open to a wide range of participants, but can also be understood as a means of restoring dignity to the places it passes through, by offering development opportunities that are proportionate to their environmental and social fragility and capable of integrating into local economic cycles without causing harm. However, such outcomes do not emerge spontaneously: they require an explicit design intention and a clear commitment from planners, managers, and governing bodies responsible for these territories over time. To clarify this point, it is necessary to consider the very meaning of *place*. We are not referring here to spaces or simple locations, but to contexts endowed with meaning, relationships, and memory, which is the sum of stories and history, which over time may have been reduced to anonymous elements or even non-places.

The increase in speed in contemporary mobility systems has gradually focused attention on departure and arrival points, relegating what lies in between to a residual dimension. At best, the journey becomes a functional interval, a necessary pause solely for the purpose of recovering energy to set off again. If deliberately pursued, cycling tourism, as a slow practice, allows this paradigm to be reversed: while maintaining the relevance of the starting and ending points of the journey, it gives central importance to the territory slowly crossed, transforming it from an indistinct mass of natural and anthropic landscape into a place to be observed, experienced, and understood. The meaning of place we want to give corrisponds to the italian word 'luogo', coming from the latin 'locus, loci', quite different from place as standardized space, like a parking lot. Getting to know an area activates a generative process that stimulates, we hope, continuous observation, care, and responsibility towards what one is travelling through. This mechanism is one of the most effective ways of protecting and enhancing places, promoting their renewed vitality. This perspective takes on particular significance in areas considered to be most disadvantaged by contemporary society, such as marginal areas: contexts that are often partially or totally excluded from public policies and unattractive for new settlements. In these cases, cycling tourism can be a strategic resource from which to initiate processes of territorial regeneration, as demonstrated by established experiences along the Danube in Austria and Germany or along the Camino de Santiago in Spain (Moscarelli 2020).

The image that best represents a long, slow project is that of a *collier of pearls*. The territories are dotted with elements of value such as landscape, cultural, and historical that can be considered individually as precious assets. However, it is their connection through a coherent cycling tourism project that completes the string, giving added value to each element. The thread that connects these pearls allows fragmented territorial narratives, interrupted over time by processes of abandonment or marginalisation, to be pieced back together. From this perspective, territorial empowering is not only about the physical possibility of reaching a place, but also about the ability to build a continuous and non-exclusive relationship between territories, people, and mobility practices, even if they are far from each other. Cycling tourism, in fact, not only makes individual points of interest or destinations accessible. While recognising the importance of points of interest, these cannot be the only pillar guiding the design of cycling routes. On the contrary, the focus shifts to the lines that generate slow movements and how these lines are experienced by those who travel along them.

Designing cycling tourism infrastructure, therefore, does not simply mean responding to users' preferences or expectations along a route. Rather, it requires critically questioning how those who travel through a territory can be guided to move beyond a purely functional or self-referential use of space, and to become attentive to the environmental, social, and cultural qualities of the places they cross. The project itself plays a central role in shaping this form of awareness, intentionally fostering the ability to perceive, interpret, and value the landscapes encountered along the way. This perspective finds historical confirmation in routes such as the Via Francigena, where it was not pre-existing urban centres that generated the route, but rather the continuous movement of pilgrims and merchants that, over time, contributed to the emergence of cities and nodes of exchange. Places such as Siena, Pavia, and Piacenza became important medieval centres of commercial and cultural life precisely because they were embedded within a network of slow and repeated passages. The linear dimension of cycling tourism, when deliberately designed with this educational intent, also enables a more balanced distribution of tourist flows. Cycling routes can act as spatial devices that allow gradual and progressive access to the territories (not the big cities) they cross, counteracting the concentration of visitors in a limited number of destinations and encouraging a wider redistribution of economic and social benefits. In this way, major urban centres and marginal areas can be connected within a single territorial system, contributing to more equitable forms of tourism development.

From this perspective, territorial empowerment through cycling tourism can be understood as the capacity to enhance the agency, resilience, and development potential of areas that are traditionally difficult to access due to morphological, infrastructural, or locational constraints. Well-designed cycling infrastructure does not automatically achieve these outcomes, but can create the conditions for local communities and territories to participate in tourism and development processes, fostering visibility, engagement, and new synergies. This approach differs fundamentally from tourism models based on attractiveness, which concentrate flows in a limited number of destinations. Such models, focused on intensive consumption, often generate marginalisation and social tensions, as evidenced by the tourist overcrowding observed in many Italian cities in recent years. Conceiving cycling tourism as a vector of territorial accessibility implies the strategic enhancement of marginal areas: territories distant from major service centres and characterised by limited access to education, healthcare, and institutional services, as well as by depopulation and youth outmigration. When integrated into deliberate planning and governance strategies, cycling tourism can contribute to mitigating, and in some cases counteracting, these dynamics of abandonment, activating local development potentials without overburdening the fragile context.

Within this framework, the rise of electric bicycle mobility is helping to redefine the conditions of territorial accessibility. E-bikes reduce the discontinuities associated with slopes, distances, and complex morphological characteristics, making areas that would otherwise be difficult to reach, such as hilltop villages, more accessible. In areas crossed by cycling paths, many of the most inaccessible places often coincide with the most marginal ones; the use of e-bikes can therefore help to narrow the gap between easily accessible areas and those traditionally excluded from cycling tourism flows. As discussed in the previous section, electric cycling mobility also broadens the audience of users potentially involved in cycling tourism. Areas that become more accessible thanks to e-bikes can benefit from this increase in flows, with positive repercussions on local economies. In particular, considering that a significant proportion of new users are elderly people, who tend to travel outside peak tourist periods, electric cycling tourism can contribute to greater stability in flows over time and to the continued sustenance of local economies.

In this context, cycling tourism and electric cycling mobility emerge as potentially effective tools for expanding territorial empowerment and reducing certain forms of spatial marginalisation. However, extending access to new territories and new types of users is not a neutral process, nor is it without consequences. Increased empowerment raises important questions in terms of environmental, infrastructural, and social impacts, which require critical reflection on how to integrate and plan electric cycling into cycling tourism contexts.

4 Is E-Slow Tourism Really Empowering and Inclusive?

In the previous sections, we discussed the concepts of inclusivity and empowerment in cycling tourism, extending to the use of electric bicycles. We explored how the introduction of electric cycling changes user engagement and the relationship between travellers and their means of transport, particularly through the introduction of external energy dependence. In addition, we analysed the potential contribution of e-bikes in making areas that are difficult to reach by traditional cycling tourism more empowered. Within this framework, the critical issues related to the integration of cycling tourism and electric cycling have so far only been mentioned in a preliminary way. The aim of this concluding section is therefore to explain and explore the main tensions and controversies that this emerging practice brings with it, highlighting how increased empowerment is neither a neutral nor consequence-free process.

It has been highlighted how electric cycling can significantly expand the pool of users potentially involved in cycling tourism. However, this inclusive potential is compared with a structural trend in the sector: the cost gap between electric bicycles and traditional muscle-powered bicycles. This economic difference tends to exclude large sections of the population who do not have the resources to sustain a high initial investment (McCurdy et al. 2024), conflicting with the spirit of inclusiveness discussed in Sect. 2. As an alternative to purchasing, many users choose to access electric cycling tourism through rental services, but this makes them dependent on the availability and organisation of such services. While this dynamic may represent an opportunity for the local economy of the areas crossed, it also introduces a form of dependency that limits the autonomy of the cycling tourist, affecting the travel experience and freedom of choice of means of transport.

Another issue linked to the expansion of the user base concerns the quality and type of cycling experience that users have during cycling tourism activities. In the case of traditional bicycles, reaching a destination is closely linked to the cyclist's level of training and physical abilities. As a result, routes located in hilly or mountainous areas are selective and not accessible to everyone. With the introduction of electric cycling, this relationship between physical ability and experience is gradually diminishing, and in some cases disappearing altogether. A study (Mitterwallner et al. 2021) has shown that changing the type of vehicle used can lead to a change in behaviour related to cycling tourism, encouraging people to overestimate their physical abilities and reducing their awareness of risk, even among less fit sections of the population. The impact of electric cycling on the cycling tourism sector could therefore translate into an increase in the potential for accidents, with consequent negative repercussions not only on individuals but on the entire community in terms of safety, emergency management, and public responsibility.

A further consequence of the expansion of the user base and reduced risk awareness concerns the environment and ecological consciousness. In particular, Mitterwallner et al. (2021) emphasises that as cyclists push themselves to go "faster, higher, further", the risks to flora and fauna increase. This observation directly echoes the reflections of Alex Langer, who proposed the principle of "*lentius, profundius, suavius*" as an alternative to our fast-paced society, arguing that without a reduction in pace, a real ecological conversion would be neither possible nor desirable (Langer 2015). The relationship between increased speed, intensified use, and environmental degradation is therefore clear, making it necessary to critically reflect on the use of electric cycling in cycling tourism. In this sense, it may be appropriate to limit or regulate access to sensitive areas or fragile natural environments, considering that electric bicycles, which are heavier and potentially more impactful, can contribute to increased erosion of trails and disturbance of wildlife. Some initiatives in this direction have already been launched in Europe. In Norway, for example, a law has been proposed to ban the use of electric bicycles in natural settings because they '*could lead to increased conflict with nature and outdoor life*' and '*increase wear and tear on trails and the natural environment*' (Ministery of Climate and Environment of Norway 2019). Similarly, in 2019, the German Alpine Club (Deutscher Alpenverein 2019) approved guidelines for the use of e-mountain bikes in mountain environments, to raise awareness of the relationship between technology, user behaviour, and ecosystem protection. However, reflection on these issues is still at an early stage and requires further study in terms of both research and regulatory measures.

Finally, another critical issue concerns territorial accessibility linked to the need for energy infrastructure. The spread of electric bicycles inevitably involves the planning of a network of charging infrastructure, which directly affects the usability of territories. To fill this gap in the literature, recent studies conducted by the authors (Pirolo et al. 2024; Pirolo and Moscarelli 2025) have proposed models for integrating slow tourism and electric cycling mobility into regional public policies. In these studies, recharging is conceived not as a simple technical break, but as an opportunity to discover the surrounding area, its cultural resources, and its natural beauty. This results in a strong responsibility for legislators, both in terms of the location of charging infrastructure and the definition of criteria for the allocation of public funds for such interventions. These choices have a direct impact on the very concept of territorial accessibility, determining which places are made effectively accessible and which remain marginalized. Only through public policies planned on a broad and consistent territorial scale is it possible to have a significant impact on increasing the accessibility of territories, including the most marginal ones.

In conclusion, the integration of electric cycling into cycling tourism shows significant potential in terms of inclusiveness and territorial empowerment, but at the same time, introduces new tensions that cannot be ignored. The critical issues that have emerged, related to access costs, user behaviour, environmental impacts, and infrastructure planning, highlight how technological innovation, if not accompanied by a conscious design and political vision, risks reproducing forms of exclusion and consumption of territories already observed in other tourism models. The issue is therefore not whether electric cycling tourism is empowering and inclusive in itself, but understanding under what conditions it can be so. Only through an integrated approach, combining design, regulation, and a culture of slowness, can electric cycling mobility contribute consistently to the objectives of slow tourism and the balanced development of territories.

Author Contributions and Acknowledgments

The authors have shared the concept of the article. P.P. supervised the entire work. L.P. wrote Sects. 1 and 4. Altogether, the authors wrote Sects. 2 and 3.

The paper is a result of the study carried out within the MOST—Sustainable Mobility National Research Center and received funding from the European Union Next-Generation EU (PIANO NAZIONALE DI RIPRESA E RESILIENZA (PNRR)—MISSIONE 4 COMPONENTE 2, INVESTIMENTO 1.4—D.D. 1033 17/06/2022, CN00000023). This chapter reflects only the authors' views and opinions; neither the European Union nor the European Commission can be considered responsible for them.

References

ADFC Bicycle Travel Analysis (2023) Cycling tourism in Germany keeps growing after the pandemic. https://pro.eurovelo.com/news/2023-03-16_adfc-bicycle-travel-analysis-2023-cycling-tourism-in-germany-is-growing-again-after-the-pandemic. Accessed 15 Oct 2023

Aprigliano V, Toro C, Rojas G, Fukushi M, Bastías I, Seriani S, da Silva WR, Peña Á, López-Quijada L (2025) The potential use of electric bicycles in the historic quarter of the seaport city of Valparaíso, Chile, through participatory mapping and focus groups supported by AI data processing. ISPRS Int J Geo Inf. https://doi.org/10.3390/ijgi14100405

Baltazar J, Vallet F, Garcia J (2022) A model for long-distance mobility with battery electric vehicles: a multi-perspective analysis. Procedia CIRP. https://doi.org/10.1016/j.procir.2022.05.259

Blanco E (2011) A social-ecological approach to voluntary environmental initiatives: the case of nature-based tourism. Policy Sci 44(1):35–52

Burns P, Bibbings L (2009) The end of tourism? Climate change and societal challenges. Twenty First Century Soc 4(1):31–51

Calzati V, de Salvo P (2017) Slow tourism: a theoretical framework. In: Slow tourism, food and cities. Routledge, pp 33–48

Cecere R, Terraferma M, Izzo F, Masiello B (2021) The influence of stakeholders in the birth stage of bike tourism networks: an exploratory study in Italy. Entrepreneurial connectivity. Springer, Singapore, pp 145–167

Ceron JP, Dubois G (2003) Tourism and sustainable development indicators: the gap between theoretical demands and practical achievements. Curr Issue Tour 6(1):54–75

Christin R (2019) Turismo di massa e usura del mondo. Eleuthera

Ciascai OR, Dezsi Ş, Rus KA (2022) Cycling tourism: a literature review to assess implications, multiple impacts, vulnerabilities, and future perspectives. Sustainability 14(15):8983

Conway D, Timms BF (2010) Re-branding alternative tourism in the Caribbean: the case for 'slow tourism.' Tour Hosp Res 10:329–344

Corti F, Iacono SD, Astolfi D, Pasetti M, Vasile A, Reatti A, Flammini A (2024) A comprehensive review of charging infrastructure for electric micromobility vehicles: technologies and challenges. Energy Rep. https://doi.org/10.1016/j.egyr.2024.06.026

Deutscher Alpenverein (2019) Sektion Oberland: Keine E-bike-Ladestationen aut Hütten. Alpinwelt 1:81. https://www.alpenverein.de/artikel/der-dav-und-das-e-mountainbike_54693750-7a48-4a5b-b350-8f410e01a804. Accessed 07 Jan 2025

Dickinson J, Lumsdon L (2010) Slow travel and tourism. Routledge

Gössling S (2000) Sustainable tourism development in developing countries: some aspects of energy use. J Sustain Tour 8:410–425. https://doi.org/10.1080/09669580008667376

Gössling S (2002) Global environmental consequences of tourism. Glob Environ Chang 12(4):283–302

Han H, Kim W, Kiatkawsin K (2017a) Emerging youth tourism: fostering young travelers' conservation intentions. J Travel Tour Mark 34(7):905–918

Han H, Meng B, Kim W (2017b) Emerging bicycle tourism and the theory of planned behavior. J Sustain Tour 25(2):292–309

Heitmann S (2011) Authenticity in tourism. Research themes for tourism. CABI, Wallingford UK, pp 45–58

Illich I (1974) Energy and equity

Kaján E, Saarinen J (2013) Tourism, climate change and adaptation: a review. Curr Issue Tour 16(2):167–195

Langer A (2015) Il viaggiatore leggero. Sellerio

Matos R (2012) Can slow tourism bring new life to alpine regions? In: The tourism and leisure industry. Routledge, pp 93–103

McCurdy A, Perry EE, Leahy JE, Coleman KJ, Doyle J, Kiewra LA, Deliyski M (2024) Gaining traction on social aspects of e-biking: a scoping review. Sustainability 16(17):7397

Ministery of Climate and Environment of Norway (2019) Amendments to the land use act (adjustments to the land boundary, etc.) access to use e-bikes in Marka. https://www.regjeringen.no/no/dokumenter/prop.-11-l-20182019/id2618254/?ch=4. Accessed 07 Jan 2025

Mitterwallner V, Steinbauer MJ, Besold A, Dreitz A, Karl M, Wachsmuth N, Audorff V (2021) Electrically assisted mountain biking: riding faster, higher, farther in natural mountain systems. J Outdoor Recreat Tour 36:100448

Mordor Intelligence (2020) E-bike market—growth. Trends, and Forecast. Morder Intelligence. https://www.mordorintelligence.it/industry-reports/e-bike-market. Accessed 07 Jan 2026

Moscarelli R (2020) Il turismo lento come occasione di pianificazione territoriale e di rinnovamento urbano: il caso del cammino di Santiago. Archivio di Studi Urbani e Regionali 129(3):152–174

Peeters P, Çakmak E, Guiver J (2024) Current issues in tourism: Mitigating climate change in sustainable tourism research. Tour Manage 100:104820

Petroman IM, Văduva L, Marin D (2021) Benefits of slow tourism for local economies. Quaestus 18:280–291

Pileri P, Moscarelli R (2021) Cycling & walking for regional development. Springer, Switzerland

Pileri P (2020) Progettare la lentezza. Linee antifragili per rigenerare l'Italia a piedi e in bici. People

Pirolo L, Belotti P, Malucelli F, Moscarelli R, Pileri P (2024) Optimal charging station location in a linear cycle path with deviations. In: International symposium on combinatorial optimization. Springer Nature Switzerland, Cham, pp 361–375

Pirolo L, Moscarelli R (2025) Integrating cycling tourism, e-mobility and local development: a spatial model to plan slow tourist paths for e-bikes. Reg Sci Policy Pract 17(12):100237

Rea E (2017) Il Po si racconta: uomini, donne, paesi, città di una Padania sconosciuta. Feltrinelli Editore

Riga DT (2025) Towards slowness: exploring human environmental experience of landscape during movement. Politecnico di Milano, Doctoral thesis. POLItesi. https://www.politesi.polimi.it/handle/10589/241497

Ritchie BW (1998) Bicycle tourism in the South Island of New Zealand: planning and management issues. Tour Manage 19(6):567–582

Rosa H (2015) Accelerazione e alienazione. Giulio Einaudi Editore

Scott D (2011) Why sustainable tourism must address climate change. J Sustain Tour 19(1):17–34

Simpson MC, Gössling S, Scott D, Hall CM, Gladin E (2008) Climate change adaptation and mitigation in the tourism sector: frameworks, tools and practices. Oxford University Press, Oxuniprint

Sun Y-Y, Faturay F, Lenzen M, Gössling S, Higham J (2024) Drivers of global tourism carbon emissions. Nat Commun 15:10384. https://doi.org/10.1038/s41467-024-54582-7

Weinert J, Ma C, Cherry C (2007) The transition to electric bikes in China: history and key reasons for rapid growth. Transportation 34(3):301–318

Sketches of a Change: Landscape as Infrastructure and Mobility as a Deal (MaaD) to Draw New Scenarios for Unexpressed Areas

Sergio Fortini and Elisa Uccellatori

Abstract The text addresses two different visions for an integrated scenario: *Landscape Metropolis* places sustainable intermodality as a premise for regeneration; *Mobility as a Deal* strengthens the social value of mobility and accessibility to unexpressed territories. Their combined action can build a system able to solve environmental, social and economic emergencies.

1 Mobility: A Human Role

In defining the principles and rules of governance of communities, planning often interprets the issue of mobility as a tool aimed at improving the conditions for economic and social development, simplifying and reducing, in fact, a primary need of the human being to an essential corollary of a productive system.

Bringing the subject back to a non-derogable faculty of human life, a decisive premise of knowledge and contamination and the directional helm of hopes, desires, fears and expectations means laying the foundations for a reversal of relations, to arrive at deciphering mobility as the backbone discipline of the territorial organization system (of soils and waters), to guide its habitability, use or abandonment.

From this perspective, the role of mobility becomes decisive for rethinking the presence of people in the territories, in contrast with the postulate that would now identify cities and large agglomerations as the future settlement. In general, vast areas of Europe and, in particular, Italy with its orographic conformation and settlement history can demonstrate how, in a historical period strongly marked by the need to regenerate rather than expand, those components capable of constituting an effective and widespread system of intermodal mobility—and reactivating physical, social, cultural and, therefore, economic networks—are present in the plots of the landscape.

S. Fortini (✉) · E. Uccellatori
Cultura della Città (società cooperativa), Ferrara, Italy
e-mail: s.fortini@culturadellacitta.it

© The Author(s), under exclusive license to Springer Nature Switzerland AG 2026

R. Moscarelli et al. (eds.), *The Future of Slow Mobility*, Research for Development,
https://doi.org/10.1007/978-3-032-23660-9_7

1.1 The Need to Build a Vision

Italy seems to have abdicated for several years the effectiveness of serious programming in the area of mobility (sustainable or otherwise), instead favouring choices in line with the political atmosphere of the moment and with easy consensus; within the political apparatus and, in several cases, of the working groups that develop the rules of regional urban plans, infrastructures are often infrastructures often come to the fore in times of crisis, almost exclusively to regenerate the muscle tone of economic activity (Spirito 2023).

Building a strategic vision therefore means trying to demonstrate that another paradigm is possible, especially in a historical period strongly marked by environmental urgency, which complements the economic one and, in many cases, the social one, already underway. The strategic vision of *Landscape Metropolis* (Fortini 2018) starts precisely from the need to tune these three urgencies within an organic reasoning, hypothesizing how, within the Italian territory, anthropic situations readable within landscape –as habitat of peculiar and rich voids of potential—can be identified rather than the ones spotted in urban agglomerations, more or less extensive, conceived as systems of "full" buildings, divided into active or abandoned ones.

Precisely in the face of an overall reasoning, underway for a couple of decades, which claims to address these urgencies under the key-word of "regeneration", it seems appropriate to build a territorial policy capable of organizing the premises for a real, non-rhetorical, regeneration, able to restore dignity to the landscape heritage that distinguishes Italy (an approach that can generally be extended to all European territories) by focusing on the potential livelihood of defined "fragile" areas in the urban planning debate—but for which the adjective "unexpressed" (Fortini 2025) is suggested—with the intention of making explicit their potential value, often historically demonstrable, then evaporated in the dynamics of contemporaneity.

Landscape Metropolis aims to cancel the city-country dichotomy, trying to design the principles of a restart that begins from landscape—be it urbanized or widespread—as a matrix of human presence and wit and as the best reticular infrastructure capable of connecting different territories in the logic of a democracy of accessibility.

1.2 Landscape Metropolis

Landscape Metropolis is a territorial policy based on the fundamental premise of using and enhancing *landscape as an infrastructure*. *Landscape Metropolis* involves a system of spaces of huge territorial extension within which the entropized dimension is abundantly exceeded by landscape.

The rediscovery of a capillary network of paths ensures a real sustainable mobility infrastructure, to be used by bike, boat, train, electric bus. Such a system becomes itself the basic prerequisite for a recovery of territories: healthy environment, social and economic repartition and recovery of meaning. The vision of *Landscape Metropolis* overcomes the dichotomy city/non-city, replacing the concept of resilience with that of anti- fragility, considering urban and territorial policies as complex systems instead of 'compact systems'.

1.2.1 A Shift of Paradigm for Unexpressed Areas

Starting from the awareness of a cultural paradigm shift, *Landscape Metropolis* proposes a strategic vision for "unexpressed area", starting from the example of the Province of Ferrara as a work in progress. Enhancing landscape as a useful infrastructure leads to a complex yet promising system of effects, which can be summarised in:

- starting a circular economy process by developing the system of resources endogenous to the territory
- strategically developing a process of significant and steady reduction in CO_2 emissions by 2030
- providing cultural changes for a healthy way of life
- contributing to protection of biodiversity
- promoting landscape and cultural heritage and generating "futurible capital"[1] (Fortini 2018)
- nurturing sense of belonging of communities and process of repopulation
- linking financing for environmental security to financing for infrastructure, with important savings for interventions (no longer episodic and activated ex-post ones)
- generating economies of scale for repairing infrastructure, urban and territorial regeneration, new working dynamics in agriculture, water regimentation, manufacture of new generation.

To organically set up a renewal of unexpressed (better than fragile) communities starting from the valorization of landscape as an infrastructure means to lay the foundations for a dynamic civil economy.

1.3 *Gradient of Magnetism for a Regeneration*

"Unexpressed places", expression coined in place of fragile areas with the aim of bringing out the present but dormant qualities of those places, retains at the same time

[1] By "futuristic capital" we mean work (conceptual and physical) aimed not only at the protection and valorisation of capital existing, but rather aimed at building the conditions for new capital (of ideas, places, goods and products).

that thrust and those hibernated resources that, if put in place, would offer opportunities for contextual development between single wallets and growth of common good.

Landscape is not only a good to protect, but a physical, social and cultural glue acting as a healthy infrastructure for daily (quality of) life. Transition thinking can therefore be expressed through a territorial narration that overcomes the classical idea of a compact city and eludes the contemporary problems of a diffused city: a "soft polycentrism" is possible; a territorial system in which *landscape as an infrastructure* becomes physical and social bond between urban and rural. These premises generate a different approach to a management model for the new mobility system.

The infrastructural continuity that landscape can guarantee becomes a condition for the realisation of a new form of urbanity, through a process of reuse of spaces and rehabilitation of dynamics of social cohesion necessary for the renewal of places: not a landscape that is urbanized but an urbanity that is "landscaped". The transport nodes between different modes of transport are intended to form the hubs of the whole system; some are already existing; others are, at the moment, systems of spaces with latent potentialities and already predisposed for re-making sense inside the total infrastructural net. Conditions are set for a strategy on the properties to be regenerated: a "gradient of magnetism" of regeneration identifies in the hubs the new polarities, capable of attracting public and private capital.

Environment becomes the engine of development, with endogenous resources to the territories, in which dynamics of economic renewal are integrated with care of environment and places and with health of people. Inside this eco-system, we are not dealing with 'simple customers', but with active *passengers* in need of a deal with territories.

2 From Service to Deal

It is precisely this social component that indicates the paradigm shift, which can influence and set a new target in mobility policies, integrating the distinctive characteristics of *MaaS* with the social and environmental logic of the unexpressed territories.

While the *Maas* model regards the provision of an end-to-end customer experience that delivers multimodal transport choices through a seamless and integrated planning, payment and ticketing interface (IET the Institute of Engineering and Technology 2024), mostly starting from the consideration of optimizing times and costs of travel within a metropolitan context, *Landscape Metropolis* requires a model in which the territory (and not only the city) is the playing frame of the intermodal network. Landscape networks, not road networks, are at the centre of the system. This fact implies that the relationship between infrastructure and user does not remain inside a dynamic of simple 'service' (such as health service) but it opens to an emotional reaction, to a sense of belonging and, why not, a pleasure (and therefore quality of life) on the journey.

This set of characteristics leads to a redefinition of the role of sustainable intermodal mobility: no longer a simple service to be provided, but a pact that citizen (whether indigenous or tourist) tightens with territory; a pact designed to transform the approach to use mobility and, therefore, to transform the impact of the system in ecological, environmental, social and economic terms.

The deal can be strengthened and enriched by the interfaces and devices envisaged by the *MaaS* model, suggesting a technological approach aimed at the social value of recovering areas in the landscape (towns, villages) and not just larger human settlements as cities and metropolises.

2.1 Mobility as a Deal (MaaD)

Mobility as a Deal (*MaaD*) is a key-expression to describe the new management behaviour under development inside *Landscape Metropolis:* the mobility system must become a pact between the communities and the local authorities.

The term "deal" assumes different declinations of meaning, able to explain the scope of the contents of this transformation and new scenarios of the social and cultural paradigm shift. Blue infrastructures and green ones collaborate in building a widespread intermodal system, capable of providing for the travel needs of people (and goods), at the same time allowing the components of which the landscape is built to provide for non-negotiable needs to combat changes climatic conditions: evaporation, transpiration, drainage, infiltration, temporary storage, run-off discharge for the blue ones; air and water purification, temperature management, flood water management and coastal defense for the green ones.

Finding value in landscape as best infrastructure also means acting on people and on behavioural change to reduce negative consequences of climate change impacts. In this conceptual scenario, the deal that mobility establishes with communities takes on a multiple social importance, in response to the need to tune the urgencies that contemporaneity places before us:

- deal as an agreement: between communities and governing forces to make "collaborative transport" and cooperation between territories, to empower social cohesion and economic opportunities;
- deal as an alliance: between mobility and environment, played on the frame of landscape;
- deal as a bargain: on the part of the user, who will be increasingly saving on transport costs; on the part of the provider of the service, who can save money in relation to accident rates, quality of air, quality of urban space, quality of landscape;
- deal as a distribution: of symmetrical and democratic potential of accessibility, for accessibility to places must have a democratic diffusion throughout the whole territories, contributing to rebirth of unexpressed areas;

- deal as a policy: brought on by the local governments in terms of choice, reducing the role of private car and linking the costs related to environmental protection to those related to the implementation of soft infrastructure;
- deal as a progress: in relation to accessibility, social cohesion quality of life and sense of belonging of communities.

3 Political Landscape

We resume the reasoning restarting from landscape: the substratum that welcomes time and space in the succession of human life, capable of dictating the rhythm of living, traveling, stopping, as well as at the same time constituting the historical deposit of traces of human industriousness, of the attempt of communities to manage the environment and bring it back to a "friendly", livable infrastructure.

As Farinelli writes, "the landscape remains, not only in the West, the form of humanity's original cognitive pact with the Earth, beyond the different degree of consciousness of individual cultures at different historical moments" (Farinelli 2024). The landscape absorbs within itself a political aspect that is as timeless as it is changeable, translating itself as a continuous theater of the interaction between man and the environment. The role of physical, social and cultural infrastructure identifies the landscape as that organic device capable of offering a system of evolving solutions to the triad of urgencies (environmental, social, economic) that distinguishes this historical moment.

3.1 Deal as Completion of Service

If moving remains an essential requirement even in the virtual era, since it is linked to the expectations and desires of human beings to find places, work, friendships, relationships, then electronic space becomes a tool of use for the unavoidable physical space, ally and gregarious of the latter in the search for meaning of people and communities.

The paradigm shift therefore does not lie in the much-anticipated death of landscape as in the capacity of the latter to rework and reinvent itself, through (and sometimes despite) human action to collectively redesign the abilities to face urgencies and the possibilities of satisfying new expectations: "Together with the idea of landscape, past generations have also transmitted to us the awareness of change, the possibility indeed necessity of its government, of its management as a function of the future of humanity as a whole: there was not a moment in the political history of the concept of landscape when such objectives did not loom on the horizon" (Farinelli 2024).

Within a *Landscape Metropolis*, the trajectories of land and water paths integrate with those of asphalt and iron and trace anthropogenic routes that belong to landscape at least as much as woodlands and wetlands. Within this territorial policy, centered on the principle of landscape as the best infrastructure, the *MaaD* model takes shape as a social evolution and integration of the *MaaS* model: the six declinations of the concept of "deal" (agreement, alliance, bargain, distribution, policy, progress) constitute, in fact, the physical and social fabric to test the evolution of the priority characters of *MaaS* (digital transport, connectivity and data technologies) transport optimization. Far from presenting itself as an alternative, the logic of *MaaD* allows us to grasp the more communitarian and democratic characteristics of *MaaS,* giving a model that is still beardless in many ways an organic direction in responding to environmental emergencies as well as social inequalities. Even if the starting concept was developed more than ten years ago, MaaS is still going through a period of adolescence, in which pilot projects follow one another as well as some successful intuitions (such as those mentioned above); but it seems to be a strong need for collaboration between public and private sector organizations to remove barriers and bring the expected benefits. Especially considering MaaS something more than an icy system of systems, but the provision of services involving real people.

Within the *MaaS* universe and in recent years, various solutions characterized by a strong social imprint are following one another, especially in European countries: from the case of *Conduent Seamless Transportation System* (a convenient mobility service via smartphone, which offers travel services tailor-made influenced by preferences) to that of *Whim* (an application that allows subscribers to have freedom of mobility without owning a personal car), from the case of *Shotl* (an app-based platform for shared and flexible on-demand transport that matches passengers heading in the same direction to a vehicle driven by a professional driver) to that of *Padam* (a turnkey Software as a Service solution for bus and city operators to develop optimized on-demand services). Through a policy based on a *MaaD* approach, similar possible solutions are technologically implemented to combine efficiency and efficiency with aspects related to accessibility to transport, access to essential facilities, education and job opportunities, all costs of public transport and shared mobility services, allowing affordable tickets.

According to the report *Public Transport and Shared Mobility* drawn up by EGUM Subgroup:

Accessibility to transport is a pre-condition for a good social cohesion and economic dynamism of a city"; at the same time, "Distances in many European suburbs are relatively short compared with other parts of the world. Mixed-used developments but also a clear densification and concentration of daily functions in attractive, walkable neighbourhoods around public transport stations and in town- and village centres have the potential to bundle mobility demand, make public transport cost-effective and increase the usability of shared-mobility options as well as walking and cycling—even in currently more car-centric suburbs—creating more liveable, inspiring, productive, and competitive places.

This condition, common to various European territories, undoubtedly and even more so belongs to the Italian reality, constituted for the great majority of its extension by a wrinkled orography dotted with towns, villages, small towns, rather than metropolises (the conception of the "metropolitan areas" acquires in Italy an administrative and governing meaning of the territories, much more than a mirror-image restitution of a densely urbanized reality). The province of Ferrara enhances these characteristics by presenting itself as a sort of green oasis within the most anthropized plain in Europe, the Po Valley.

3.2 *A Scenario in Progress: Ferrara Idropolitana (Hydropolitan Ferrara)*

Within the approximately 2600 km^2 of its extension, the anthropized parts cover a laughable percentage compared to a landscape sprayed by rivers and canals which presents itself as a large infrastructure, organized over the centuries for the management of the relationship between land and water.

The territorial policy of *Landscape Metropolis* therefore takes shape from these landscape scenarios and orients them towards regeneration—and therefore towards the repopulation of areas which, if made accessible and reached by services, would become very attractive as places with a high quality of life—making use of a innovative conceptual development: the MaaD principle, precisely. The most ambitious project that links the strategic vision of *Landscape Metropolis* to the conceptual elaboration of *MaaD* is developed from Ferrara, the capital city of the province, and is called *Ferrara Idropolitana* (*Hydropolitan Ferrara*), for which a pre-feasibility study was carried out under the coordination of the transport agency of the Emilia Romagna region, TPER.

3.2.1 Local Public Treasure

The peculiar orography of the soil and the presence of a waterway for each cardinal point (Canale Boicelli, Po di Volano, Po di Primaro, Canale di Burana) give the municipality of Ferrara a decisive infrastructural potential in the dynamics of proximity transport, in home-school and home-work trips and, in general, for all trips that can characterize the daily life of a city.

For this reason, intermodal Local Public Transport (LPT) has been hypothesized, within which a network of landing places is integrated with bus transport and the extended cycle network. Within a UNESCO city-territory like Ferrara, the possibility of guaranteeing an intermodal system partly based on waterways constitutes suggestive and immediate added value even towards those slow tourism on which urban policies have been focusing for some time, with significantly growing numbers.

Sustainable mobility in urban areas represents the key to understanding the reduction of CO_2 and fine dust emissions, the decongestion of traffic and the reduction of accidents (with consequent optimization of public spending). It also constitutes the tool for daily monitoring of the landscape-infrastructure and for its management, especially with regards to hydrogeological instability, the protection of green areas, soil and river areas. Furthermore, in a city rich in history such as Ferrara, poised between a relatively high average age and a young university population of around 30,000 units, the opportunity to switch part of the traffic to an intermodal land–water system which, at the moment, in Italy it is not yet present (apart from the Venetian case which however is based on the land/Mestre-water/Venice dichotomy) would mean generating territorial marketing and important related activities.

3.3 Accessible Territories

The intermodal structure of mobility is transformed into a pact between managers and users, according to the declinations of the principles of *MaaD*, allowing maximum accessibility between urban center and peri-urban areas and an inverse proportion between increase of intermodal flows (sustainable) and decrease in the use of private car.

At the same time, the intermodality of the system and its capillarity imply an integrated vision of the different infrastructures and a coordination of the various modes of transport both reachable with technological instrumentation in *MaaS* logic. This system must stimulate a reversal of the dynamics with which Public Transport is undergoing, over the years, a process of penalization underlying the simplistic reasoning that places it among those services in constant economic loss and, therefore, no longer sustainable; the proposed articulation is aimed at demonstrating how the interpolated action between intermodality and technology, between sustainability of the network and (measurable) benefits for the community in terms of accessibility to places, ecosystem services and quality of living spaces and finally, more generally, between *MaaD* and *MaaS*.

The same approach as Ferrara Idropolitana can be extended, within the province of Ferrara, into other areas characterized by the presence of water as well as the public bus transport service, the railway and cycle paths, laying the foundations for the recovery of unexpressed areas and in the process of depopulation.

4 Conclusion in Progress

The strategic vision of *Landscape Metropolis* brings together the potential of intermodal mobility as a premise for the regeneration of places and territories.

Its political declination oriented to the principle of *Mobility as a Deal*—intertwined with a management and functional optimization through the appropriate technological devices, in *MaaS* logic—can generate effective implications in terms of quality of life, ecosystem services and the very health of the inhabitants, with economic repercussions (also in terms of territorial marketing) capable of reversing the marked trends towards an increase in prices to the point of being able to hypothesize, as a limit to aim for, a elimination of the same and widespread free public mobility, complicit in the environmental improvement and the state of places as well as the increase in value of all those spaces (open or volumes) that are crossed by that mobility network.

Moreover, the investment itself in public fleets and landscape infrastructures by the authorities (combined with facilitation policies especially with regard to the most marginal nuclei) would also act as an incentive to change for the entire population. Metropolitan densification is partly the result of an old capitalism, centered on extractive and aggressive logic towards the environment and landscapes. Perhaps still far from subverting capitalism in its most intimate structure, we can however change some of its defining connotations for a repositioning of the human being at the center of the system. This transformation also passes through a redistribution of the presence of man oriented towards the re-signification of all those unexpressed areas in the process of depopulation—and, therefore, in constant loss of specific weight and meaning until they become places that do not count (Rodriguez-Pose)—but capable, if reinserted into an effective system of intermodal mobility and personal services, of reactivating their landscape qualities and constituting, through a system of countries, the backbone of a Country and "futurible capital" for the next generations.

References

Farinelli F (2024) Il paesaggio che ci riguarda. Touring Servizi, Milano

Fortini S (2018) Metropoli di Paesaggio. In: Paper presented at XXI SIU national symposium of Italian Society of Urban Planners, Florence, 6–8 June 2018

Fortini S, (2025) APPIFarm Fiscaglia: the restart of "unexpressed territories" CDS Cultura Edizioni. https://www.cdscultura.com. Accessed 5 Feb 2025

IET the Institute of Engineering and Technology (2019) IET transportation series. https://www.theiet.org. Accessed 27 Jul 2025

Spirito P (2023) The assassination of transportation planning: horizons for a reconstruction. In: Territori Economie Società Istituzioni (TESI) 12. https://www.unimercatorum.it/ricerca/tesi-e-temi

Planning and Designing E-Mobility

Data-Driven Design of Human-in-the-Loop Assistance Systems for Light Electric Vehicles

Jessica Leoni and Mara Tanelli

Abstract Advanced assistance systems are crucial for enhancing light electric vehicles safety, energy efficiency, and user experience. Accordingly, this Chapter investigates the development of functionalities that jointly monitor the urban environment, the road conditions, and the vehicle dynamics to inform safety measures such as rider alerts and adaptive speed control. Then, it also explores different strategies to promote rider experience by considering rider physiological data to find the best trade-off between electric and human power, minimize battery discharge, and maximize riding enjoyment. Last, it discusses the customization of the riders' experience by considering rider preferences, habits, and driving styles. For each of these core areas, we will analyze the strengths and limitations of existing literature approaches and presents novel solutions that we have recently designed to address the identified gaps.

1 Towards a More Sustainable and Inclusive Urban Mobility

In recent decades, the pressing impact of global warming has demanded concrete countermeasures (Masson-Delmotte et al. 2022). Given the evidence that, in recent years, road transport accounts for a stable share of 37% of global energy-related CO_2 emissions, policies like the European Green Deal (Haas and Sander 2020) and China's Integrated Policy Assessment Model (Kejun et al. 2021) aiming for zero-emission transport by 2050 have been promoted. As a consequence, research in alternatives to private cars has spread (Ewert et al. 2020). Considering urban micromobility, light electric vehicles (LEVs), such as e-scooters and e-bikes, prove to be one of the most promising solutions (European Commission 2020). In fact, despite

J. Leoni (✉) · M. Tanelli
Dipartimento di Elettronica, Informazione e Bioingegneria, Politecnico di Milano, Milan, Italy
e-mail: jessica.leoni@polimi.it

M. Tanelli
e-mail: mara.tanelli@polimi.it

© The Author(s), under exclusive license to Springer Nature Switzerland AG 2026
R. Moscarelli et al. (eds.), *The Future of Slow Mobility*, Research for Development,
https://doi.org/10.1007/978-3-032-23660-9_8

being sustainable (Campisi et al. 2020), they are attractive, user-friendly, license-free (Teixeira et al. 2023), increasingly affordable thanks to battery advances, and cheaper to operate and maintain (Michelangelo-Gulino et al. 2022). However, in the urban environment, despite the growing presence of LEV, large-scale uptake is still hindered. This is because current policies often overlook that changing mobility patterns requires not only promoting sustainable travel options but also transforming mobility culture and addressing users' needs, preferences, and fears (Sopjani et al. 2019; Van Bree et al. 2010). Therefore, recent research focused on end-users' feedback to better understand and thus overcome these barriers, identifying two primary limitations. First, safety concerns, especially for e-scooters, related to the perceived high injury risk (Teixeira et al. 2023). Second, mainly for e-bikes, the absence of adaptive control. In fact, current models rely on fixed assistance levels, forcing extra effort, especially on poor surfaces, and also being unaware of specific rider needs (Christoforou et al. 2021).

1.1 Human-in-the-Loop Control Context-Aware System Architecture

Thus, previous analyses show LEVs are promising for sustainable urban mobility, but safety and fatigue must be addressed. However, designing effective human-in-the-loop, context-aware control systems is complex, requiring consideration of factors often overlooked in conventional vehicle controls. Key variables are road quality and rider characteristics. In fact, the first affects vehicle vibrations, which can cause loss of control due to the small tire-road contact area. Regarding the second, factors such as rider mass, often exceeding the vehicle's, influences stability by affecting vertical forces. Adaptive fatigue control must consider user mass, psychophysiological state, and road conditions. Thus, a human-in-the-loop, context-aware approach is essential not only to address safety and fatigue-related issues but also to promote condition-based maintenance, as these same factors are also key contributors to vehicle usage (De Bortoli and Christoforou 2020).

To move from a traditional design towards a human-in-the-loop and context-aware control system architecture, we herein present and discuss the schema depicted in Fig. 1. At a high level, it is composed of four primary blocks: two related to the rider, one related to the vehicle, and one to the controller. Embedding rider and context variability in models lets existing controllers perform better without redesign. In contrast, current controllers, e.g., location-based speed reduction for e-scooters (Brankovic et al. 2015; Leoni et al. 2024) or fatigue-based support for e-bikes (Corno et al. 2014; Meyer et al. 2015), degrade in real-world use because they do not account for rider or terrain variability.

In light of these considerations, Sect. 2 details adaptable rider behavior models; Sect. 3 covers parametric vehicle modeling including rider mass and road quality; Sect. 4 summarizes and outlines open challenges.

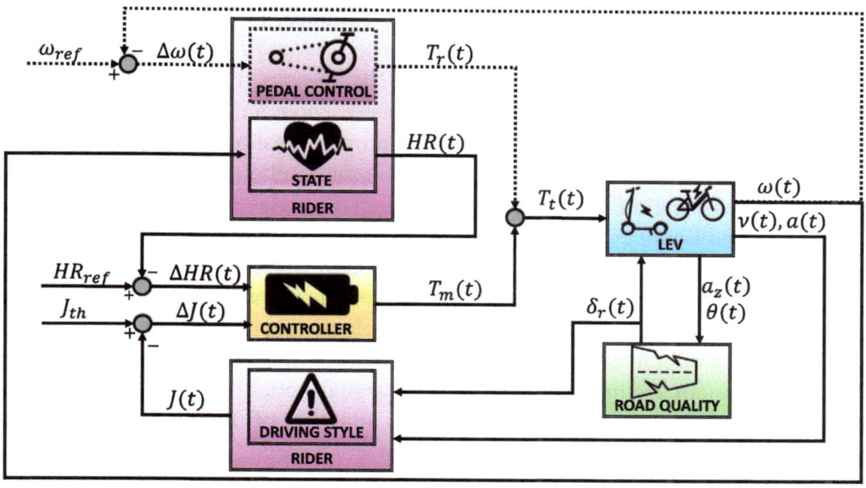

Fig. 1 Proposed human-in-the-loop, context-aware control architecture. The dotted section applies only to e-bikes. The variables are defined as follows: ω is the cadence, v the vehicle speed, a the acceleration vector, composed of lateral a_y, longitudinal a_x, and vertical a_z components; θ is the pitch rate, T_t is the total torque applied to the vehicle, which is equal to the motor torque T_m for e-scooters, and the sum of motor torque and rider torque T_r fore-bikes. HR represents the heart rate of the rider. Last, δ_r is the road friction coefficient, and J is the vector of driving style indices

2 Human Behavior Estimation

Moving to the user block, it must be noticed that it can consist of several sub-components, based on the behaviors of interest in the target application. Usually, driving style and psychophysiological state are modeled. However, when considering e-bikes, the rider's pedal control action can also be taken into account.

2.1 *Driving Style*

The literature provides several examples of how to estimate driving style risk for traditional vehicles based on sensor data. These approaches can be categorized into two main groups: machine learning-based and index-based methods. An example of the first category is a recent study (Brankovic et al. 2015), which introduces a machine learning framework to estimate driving style-related risk in real-time; however, it leverages information collected by a lidar sensor, which is difficult to implement in LEVs, particularly within a sharing paradigm. In addition, approaches such as (Brankovic et al. 2015) assess riding style using cheaper sensors, such as IMU and GPS, but leverage deep neural networks, which raises certification issues and limits the feasibility of these solutions.

Index-based methods are also effective. For instance, in Knowles et al. (2012), driving style is assessed using a jerk-based metric $\Gamma = \sigma_J/\overline{J}$, with static thresholds to classify risk. Similarly, Gelmini et al. (2020) propose dynamic-aware indexes for braking, acceleration, lane changes, and speed. Though designed for cars, these approaches can be adapted to LEVs given that proper threshold tuning has been performed to account for the different rider-vehicle mass relation (Felipe et al. 2015). Thus, taking inspiration from the approaches in Knowles et al. (2012), Gelmini et al. (2020) and Leoni et al. (2023) we define four indexes to estimate rider driving style, which have been properly tailored to address the context of LEV. Specifically, those indexes rely on three fundamental vehicle dynamics represented by the accelerations $a_x(t)$, $a_y(t)$, and $a_z(t)$, and on the velocity signal $v(t)$. Since frequency-domain features effectively capture vehicle dynamics in all three spatial directions, while time-domain features suit the velocity signal better, we based our indexes on the area under the Power Spectral Density function of the three accelerations, namely PSD_{a_x}, PSD_{a_y}, and PSD_{a_z}, and the average speed $\overline{v}$. By real-time extracting these features using an ad-hoc fine-tuned rolling window, following the approach in Gelmini et al. (2020), whole-trip indexes describing the longitudinal, lateral, vertical, and speed-related dynamics can be derived by summing feature values over time whenever they exceed a tailored threshold, which we ad-hoc set for addressing LEVs dynamics. Specifically, for each time window w, the computed set of indexes $J(t) = \{\overline{v}w(t), PSDa_x, w(t), PSDa_y, w(t), PSDa_z, w(t)\}$, can be computed as:

$$J_v(t) = \begin{cases} 0 & \text{if } \overline{v}_w(t) < v_{th} \\ \overline{v}_w(t) - v_{th} & \text{otherwise.} \end{cases} \tag{1a}$$

$$J_{PSD_{a_x}}(t) = \begin{cases} 0 & \text{if } PSD_{a_x,w}(t) < psdx_{th} \\ PSD_{a_x,w}(t) & \text{otherwise.} \end{cases} \tag{1b}$$

$$J_{PSD_{a_y}}(t) = \begin{cases} 0 & \text{if } PSD_{a_y,w}(t) < psdy_{th} \\ PSD_{a_y,w}(t) & \text{otherwise.} \end{cases} \tag{1c}$$

$$J_{PSD_{a_z}}(t) = \begin{cases} 0 & \text{if } PSD_{a_z,w}(t) < psdz_{th} \\ PSD_{a_z,w}(t) & \text{otherwise.} \end{cases} \tag{1d}$$

where the threshold array for these indexes is denoted as $J_{th} = v_{th}, psdx_{th}, psdy_{th}, psdz_{th}$. Then, the integral indexes for an entire trip are computed as the sum of the squared adjusted feature values, normalized by the trip length d. For the ith metric, this is given by:

$$J_i = \frac{1}{d}\sum_{w=1}^{W} m(w)^2, \tag{2}$$

where W represents the whole number of extracted windows over the trip. To evaluate the proposed framework, data collected in an extensive campaign was considered. It involves a diverse group of 18 volunteers with different riding experiences. Each

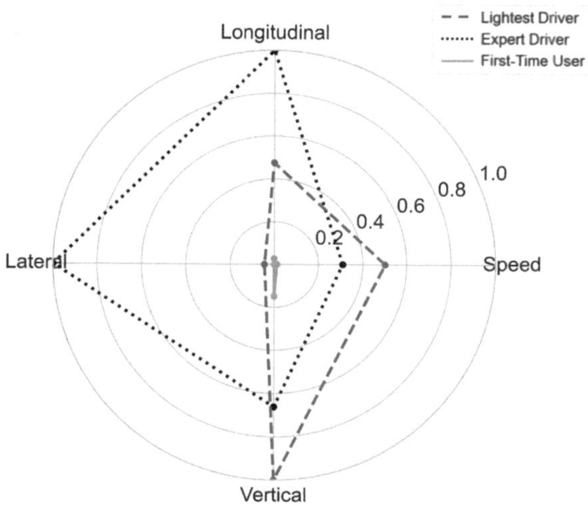

Fig. 2 Whole trip style indexes. This Figure reports the integral indexes computed considering three emblematic trials

volunteer rode a predefined 1.7 km asphalted route. Despite the lack of a real-time ground truth for driving style, the aggregated indexes provide consistent results with a priori knowledge of the riders' attitude. Specifically, we can notice that more experienced riders tend to exhibit extreme driving styles, i.e., attempting to complete the route in the shortest time possible. In contrast, first-time users adopt cautious behavior, especially during lateral maneuvers. Figure 2 shows, using a radar chart, the results obtained for three representative trials performed by an expert rider who drives aggressively, a first-time user who drives calmly, and the lightest rider who was confident with the vehicle and drives in a normal style. Specifically, each polygon corresponds to one of the trials, and each vertex represents the integral index value obtained for the longitudinal, lateral, vertical, and speed index, respectively. We can notice that, consistently with what is expected, the vertical dynamics and speed indexes are highest for the lightest rider; on the other hand, considering the expert driver that drive aggressively, the lateral dynamic index is instead significantly higher for the experienced rider due to aggressive lane changes, and the longitudinal dynamics index is amplified due to abrupt acceleration and braking maneuvers.

2.2 *Psychophysiological State*

Besides driving style, the rider's psychophysiological state is relevant to designing controls that enforce safety and reduce fatigue. However, real-time estimation of rider fatigue is challenging due to individual differences like age, fitness, and health. While complex muscle fatigue models exist (Wan et al. 2014), they are impractical for real use. Luckily, research proves that heart rate is a reliable fatigue proxy, with fatigue occurring when heart rate exceeds 50–60% of the maximum, which is commonly approximated using Tanaka's formula (Sarzynski et al. 2013):

$$HR_{max} = 220 - age \tag{3}$$

From this value, the mentioned threshold can be computed as 50%-60% of HR_{max}. Building on this, various models have been developed to describe the dynamic relationship between heart rate and cycling effort, which can be categorized into two groups: linear and nonlinear. Linear models, as that presented in Lefever et al. (2014), despite being often preferred due to their lower complexity, sometimes fail to represent more complex heart dynamics. On the other hand, nonlinear models significantly complicate both the controller design and the computation of the control action. In this context, a promising solution is then given by the Hammerstein model (Su et al. 2007), which integrates the nonlinear dynamics through a static function applied to the input variable. This function between the heart rate HR and the rider power P_r can be both linear and nonlinear and is formulated as:

$$\mathcal{N}_{lin} : HR = a_1 + a_2 \cdot P_r \tag{4a}$$

$$\mathcal{N}_{nonlin} : HR = a_3 + a_4 \cdot P_r^2 \tag{4b}$$

where the coefficients a_1, a_2, a_3 and a_4 must be estimated by a least squares approach from data collected in the first part of a trial, during the acceleration region, without any assistance from the e-bike motor. The output of this nonlinear function feeds a linear model that predicts heart rate changes from a reference HR_0 in response to cycling effort. This linear model is composed two parts for short- and long-term fatigue. However, since e-bikes are typically used for short trips, only the short-term response is relevant, modeled by this second-order linear function:

$$G_{HR}(s) = \frac{\mu_{HR}}{(1 + sT_{HR,1})(1 + sT_{HR,2})} \tag{5}$$

where μ_{HR} represents the gain for the steady state displacement of the heart rate from its baseline value, while $T_{HR,1}$ and $T_{HR,2}$ influence the shape and the duration of the transient to reach the steady state. Once the coefficients of this equation have been estimated, the heart rate can be modeled as:

$$HR(t) = HR_0 + G(s) \cdot \mathcal{N}(P_r(t)) \tag{6}$$

where HR_0 is the baseline heart rate of the rider, perceived in a relaxed condition before cycling. This separation of nonlinear input and linear dynamics simplifies control design and provides a more accurate system representation. When applied to several case studies, Hammerstein models have shown effective results, and often prove that a linear static function is enough to represent the system behavior in normal ride, while a nonlinear input function becomes relevant in the case of intense cycling (Corno et al. 2014).

To prove the effectiveness of this approach, we use it to model the heart rate dynamics of the rider involved in our e-bike case study, considering the data used also for validating the vehicle dynamics model presented in Sect. 3. From this data, we first estimate the static relation between rider power and heart rate. To this end, we use both linear and quadratic functions, according to the formulations reported in Eqs. 4a and 4b, respectively. Table 1 reports the coefficients for the two equations estimated by applying a least squares error optimization to the data of each rider separately, along with the corresponding $nRMSE$ and the R^2 score between actual and predicted heart rate. Figure 3 shows the results for the heavy rider, which aligns with those of the light one. As expected from the literature, given that no intense riding trial has been performed, in both cases the fitted linear model provides a better fit.

Once the static response has been estimated, the coefficients in the linear model described by Eq. 6 must also be identified. Although the gain value μ_{HR} can easily be derived as the slope in the linear function previously identified, and the baseline value of the heart rate of the rider HR_0, equal to $81bpm$ for the light rider and $72bpm$ for the heavy rider, can be measured during a rest state condition, identifying the time

Table 1 Estimated input function

Rider	Model	Evalauation metric	
		nRMSE	R^2 score
35 kg	$\mathcal{N}_{lin} : HR = 94.787 + 1.007 \cdot P_r$	0.185	0.584
	$\mathcal{N}_{nonlin} : HR = 109.672 + 0.012 \cdot P_r^2$	0.211	0.458
90 kg	$\mathcal{N}_{lin} : HR = 77.651 + 0.307 \cdot P_r$	0.125	0.667
	$\mathcal{N}_{nonlin} : HR = 85.979 + 0.002 \cdot P_r^2$	0.138	0.600

This table presents the estimated coefficients and corresponding evaluation metrics for both the linear and nonlinear input functions, considering the two riders involved in the e-bike case study

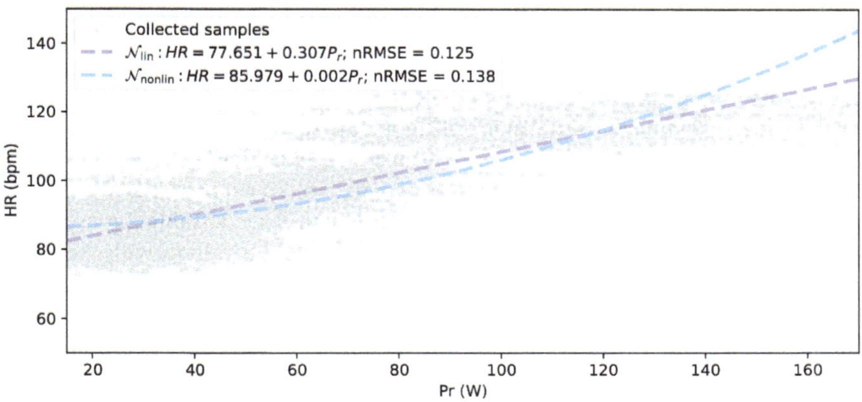

Fig. 3 Estimated input function. This figure shows the estimated heart rate versus power relation for the heavier rider, approximated using the linear and the quadratic functions

constants for the two poles is challenging. In Corno et al. (2014), these were estimated via frequency response analysis using a continuously varying transmission, which, however, is not available in most of the on-the-market e-bikes. Thus, we assume that the heart rate transients remain similar across individuals, and we adopt the values estimated in the referenced paper, which have been $T_{HR,1} = 22s$ and $T_{HR,2} = 55s$. Of course, this assumption represents a limitation that should be addressed in future research.

2.3 Pedaling Control Action

Last, specifically for e-bikes, it may also be valuable to model the rider's pedaling control action. A widely accepted model in the literature is the one proposed by Corno et al. (2014), which represents the rider as a pedaling cadence controller that compares the current cadence with the desired one and accordingly modulates the torque on the pedals. Specifically, it is defined as a linear time-invariant second-order controller:

$$R_{\omega_p}(s) = \mu_{rid} \frac{(1 + s\tau_{rid})}{(1 + sT_{rid})^2} \tag{7}$$

where μ_{rid} is the gain, τ_{rid} is the time constant of the zero, and T_{rid} is the time constant of the two coincident poles. Although this approach offers an indirect measure of the rider's fitness level and style preferences, it has some limitations. The main one is that the model parameters must be estimated for each new rider. However, in the presented study, to derive these coefficients for a single rider, multiple experiments were conducted where the e-bike is accelerated to a constant velocity starting from standstill with a fixed transmission ratio. However, this procedure is long and demanding; thus, it can not be applied in real-world scenarios.

To this end, in our work (Leoni et al. 2025), we design an online identification strategy to make this approach capable of real-time adapting to new riders. This method still leverages the fixed transfer function proposed in Corno et al. (2014) but considers the gain, and the zero and the poles time constant as adaptive coefficients. Thus, we estimate these values using a least squares optimization approach by considering the first $30s$ of a new ride, since we prove that this duration was the shortest effective one for achieving robust parameters. To evaluate our framework, we consider the trials in the e-bike dataset described in Sect. 3. Table 2 presents the estimated coefficients, aggregated by rider and road quality, along with the corresponding average $nRMSE$ between measured and simulated rider torque $T_r(t)$, in the respective trials. In addition, Fig. 5 illustrates the whole distribution for the estimated parameters, highlighting significant variations across riders and reinforcing the need for online estimation. Specifically, it is possible to notice that lighter riders exhibit lower gain, and their control action is less affected by changes in terrain quality. To further demonstrate the model's effectiveness, Fig. 4 compares the pre-

Table 2 Estimated discrepancy correction coefficients for the rider behavior model

Rider	Road quality	Parameters			Performance nRMSE
		μ_{rid}	τ_{rid}	T_{rid}	
35 kg	Good	6.05	0.94	1.34	0.56
	Medium	9.94	0.96	1.23	0.98
	Bad	9.06	0.91	0.73	0.58
90 kg	Good	23.05	0.99	1.05	0.76
	Medium	26.70	0.95	1.51	0.88
	Bad	24.07	0.97	0.34	0.73

dicted rider torque $T_r(t)$ and vehicle speed $v(t)$ for the heavy rider on asphalt using the parametric model estimated according to the presented method, using a model estimated for the same rider but on gravel, and using a model estimated on asphalt but for the light rider. All trials are conducted with $\kappa = 0$ and a reference speed of $12km/h$. Results show that using a model from a different rider significantly impacts performance, whereas using a model estimated on a different terrain has minimal effect. This suggests that rider-specific coefficients can be estimated at the start of the ride to adapt the model to a new user, but can be considered as constant throughout, as simulation results confirm that rider behavior remains relatively stable despite changes in road quality.

3 Parametric Models for Personalized Assistance

For what concerns vehicle model, the longitudinal dynamics of LEVs is often modeled based on equations derived from the dynamic equilibrium of forces outlined in Radrizzani et al. (2023). Assuming a negligible road slope, the system follows:

$$M\dot{v} = -F_{cv} - Fb + \frac{T_{tot}}{r} \tag{8}$$

where:

- $M\dot{v}$ represents the inertial force of the e-bike, where $M = M_{ref}$ is the total mass. Here, $M_{ref} = M_b + M_r$ accounts for the vehicle mass and a reference rider mass.
- F_{cv} is the coasting down force, defined as:

$$F_{cv} = \alpha v^2 + \beta v + Mg\gamma \tag{9}$$

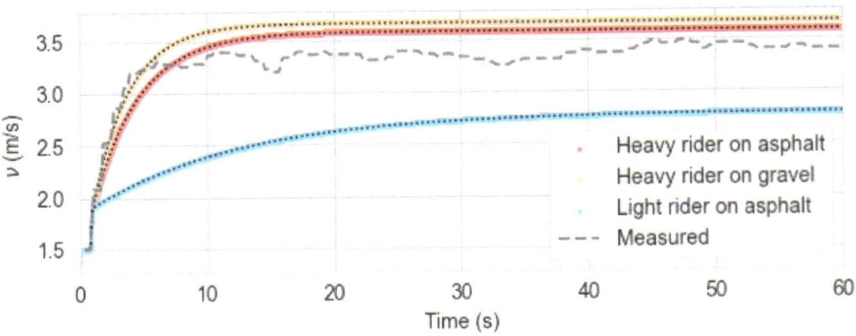

(a) Vehicle Speed

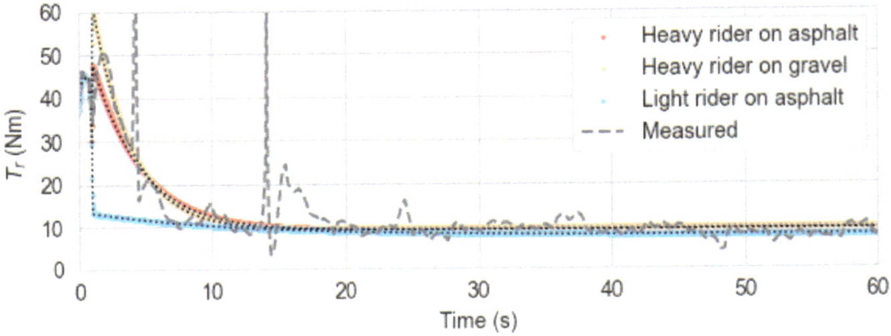

(b) Rider Torque

Fig. 4 Speed and torque predictions. This figure shows the predicted vehicle speed and rider torque for a trial performed on the e-bike by the heavy rider with $\kappa = 0$ and a reference speed of 12 km/h on asphalt. It compares the predictions of the controller trained on this trial to those of controllers trained on the same trial but performed by a lighter rider and by the same rider but on gravel

so that:

- $\alpha = \frac{1}{2}\rho C_x A$ is the aerodynamic drag; ρ is the density of the air, A the front area of the vehicle and the rider and C_x the drag coefficient.
- $\beta = D_\upsilon$, is the viscous drag coefficient.
- $Mg\gamma$, considering $\gamma = C_r$ represents the rolling resistance.

- $\frac{T_{tot}}{r}$ is the traction force, where $T_{tot} = T_r + T_m$ is the sum of rider and motor torques, and r is the wheel radius.
- Fb is the braking force.

As explained in Corno et al. (2016), these coefficients can be identified by ad-hoc performed coasting down experiments. While effective, this model does not account for rider variability and road conditions, as its coefficients are typically identified for a

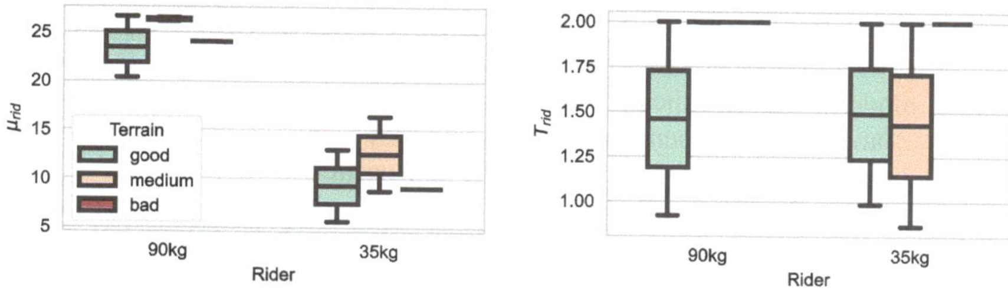

(a) Estimated μ_{rid} by rider (b) Estimated T_{rid} by terrain

Fig. 5 Parameters estimated for rider behavior model. This figure shows the distribution of the estimated μ_r and T_{rid} from the trials in the *road quality* dataset

specific rider and surface. Thus, to make this fixed model context- and human-aware, our suggestion is to reframe this dynamical equation to be parameteric. To this end, specific coefficients have been introduced, which improve its accuracy in resembling the actual vehicle dynamics (Leoni et al. 2025). The two included coefficients, δ_m and δ_r, account for changes in rider mass and road quality, so that:

- $M = M_{ref} \cdot \delta_m$, with $M_{ref} = M_b + M_r$, where δ_m compensates for deviations from the reference mass.
- The rolling resistance term is modified as $Mg\gamma$, where $\gamma = C_r + \delta_r$ and δ_r corrects the nominal rolling resistance coefficient.

While δ_m can be estimated once, at the beginning, and then considered as constant, δ_r varies dynamically during the trip. Accordingly, for each new trial, δ_m and the initial δ_r can be estimated by applying a recursive least squares approach to the data collected within the first 30 s that exceeds a speed of 5 km/h and with a positive cadence. Then, δ_r should be continuously updated during the ride, either by applying recursive least squares, though computationally expensive, or by classifying the road quality into a predefined set of categories and using a scheduled value of δ_r for each class.

To perform road quality classification, several approaches have been presented in the literature. Most of them are referred to the automotive context and rely on a standard or revised version of the traditional International Roughness Index (IRI) (Sayers et al. 1986). However, the IRI has been demonstrated to be poorly applicable when considering LEVs, since they are characterized by low travel speed and light damping (Cafiso et al. 2022). Consequently, alternative approaches can be investigated. Considering traditional vehicles, road surface quality estimation has also been addressed in the literature by proposing model-based approaches that rely on vehicle vibrations, typically measured using a simple setup consisting of a tri-axial accelerometer and GPS (Yeganeh et al. 2019). This research encompasses both simple model-based approaches for two-wheeled vehicles (Gorges et al. 2018) and four-wheeled vehicles (Hassen et al. 2019), along with advanced techniques that utilize sliding mode observers (Rath et al. 2014) and Kalman filters (Dhole et al. 2019). However, these model-based methods have limited applicability to LEVs, as they rely on a precise model to describe the LEV's vertical dynamics, which is a challenging task due to the complex, user-dependent behavior and the absence of suspension systems. Thus, attention has shifted toward model-free approaches, which relies on machine learning and deep learning techniques that are showing promising results (Yousefzadeh et al. 2010; Nguyen et al. 2019). However, most of these methods rely on information gathered from expensive sensors, such as cameras, and leverage unexplainable approaches, like deep neural networks, which hinder implementation and certification Leoni et al. (2022, 2020). Given the above considerations, in (Leoni et al. 2023) and (Leoni et al. 2025) we present a machine learning-based approach for real-time road quality classification that relies on features extracted from a simple sensor setup including an IMU. Specifically, vertical acceleration, $a_z(t)$, and pitch rate, $\theta(t)$, prove to be the most effective for reconstructing road quality, especially when considering frequency-domain features rather than time-domain ones (Hu 2015). Specifically, the area under the Power Spectral Density (PSD) of these signals is one of the most suitable metrics.

3.1 Parametric Vehicle Model Validation

To reliably validate the effectiveness of the proposed approach, it is key to rely on data collected in dedicated experimental campaigns. To this end, a robust campaign should account for riders of different masses, riding at different speeds over terrains of different quality. To validate the parametric vehicle model presented, we considered data collected during a campaign involving two riders weighing $35kg$ and $90kg$ in 36 trials, each $300m$ long. Two reference speeds of $12km/h$ and $20km/h$ have been considered. For each one, each rider performed six trials on asphalt, six on gravel, and six on cobblestones with no assistance ($\kappa = 0$) factor. For both vehicles, each trial was conducted in the same way: the rider was asked to reach and maintain the target speed and keep it constant until the end, traveling as straight as possible on the target surface. First, following the approach in (Corno et al. (2016)), we determine the coasting down coefficients (α, β, and γ) from specifically designed experiments conducted by a 90 kg rider on an asphalt surface. This scenario represents the baseline model, where $\delta_m = 1$ and $\delta_r = 0$. Under these conditions, we obtain the fixed model parameters $\alpha = 0.265$, $\beta = 0.475$, and $\gamma = 0.008$. Then, we estimated the coefficients for each trial. Figure 6 and Table 3 summarize the obtained results, revealing that the estimated δ_m was closer to 1 for the heavy rider, while it ranged from 0.47 to 0.51 for the light rider, aligning with their mass differences. Similarly, δ_r correlates with road quality, with values close to 0 for trials on asphalt, approximately 0.15 for gravel, and around 0.28 for cobblestones. Based on these findings, we decided to update δ_r using these scheduled values based on the predictions of the road quality classifier. Figure 7 directly compares the difference between the estimates of the fixed and parametric models for two trials performed on cobblestones at 12 km/h with $\kappa = 0$

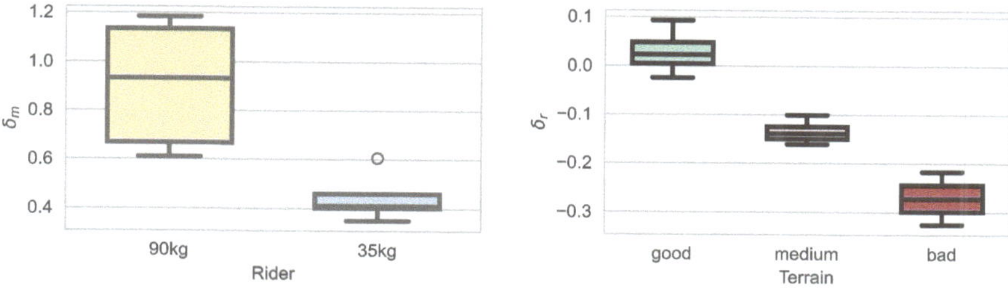

(a) Estimated δ_m by rider (b) Estimated δ_r by terrain

Fig. 6 Estimated discrepancy correction parameters. This figure shows the distribution of the estimated δ_m and δ_r from the trials in the e-bike dataset

Table 3 Estimated discrepancy correction coefficients for the vehicle dynamic model

Rider	Road quality	Parameters		Performance nRMSE
		δ_m	δ_r	
35 kg	Good	0.46	0.02	0.03
	Medium	0.49	0.11	0.15
	Bad	0.41	0.25	0.13
90 kg	Good	1.01	0.02	0.07
	Medium	1.18	0.15	0.23
	Bad	0.70	0.32	0.20

by the two riders. To update δ_r, we adopt the scheduled values-based approach described in both trials. The results show that the fixed model exhibits higher errors, particularly for the light rider, due to inaccuracies in mass estimation and road surface variations. Even for the heavy rider, the fixed parameters fail to accurately capture the system dynamics. Also, it can be noticed that variations in rider mass primarily affect the transient phase; on the other hand, changes in road quality impact steady-state performance, highlighting the necessity of real-time updates for δ_r.

4 Future Directions and Challenges

This chapter presents a human-in-the-loop, context-aware control system for LEVs. Its key novelty consists of including modular parametric models that account for rider mass, psychophysiological state, driving style, and road conditions, factors often neglected but critical for LEVs. Experimental results demonstrate the benefits of this approach. Future research will focus on adapting heart rate dynamics

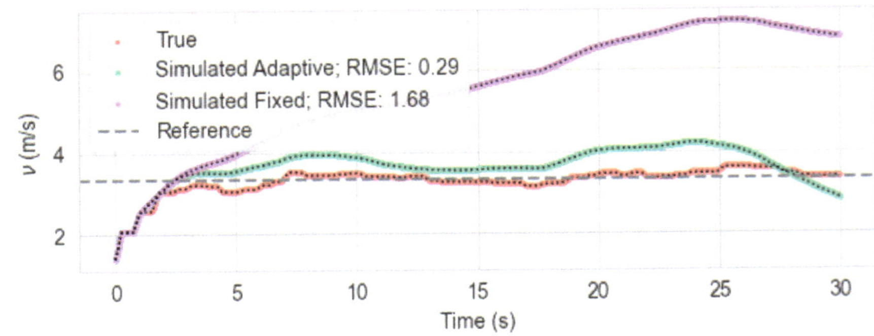

(a) 90kg rider

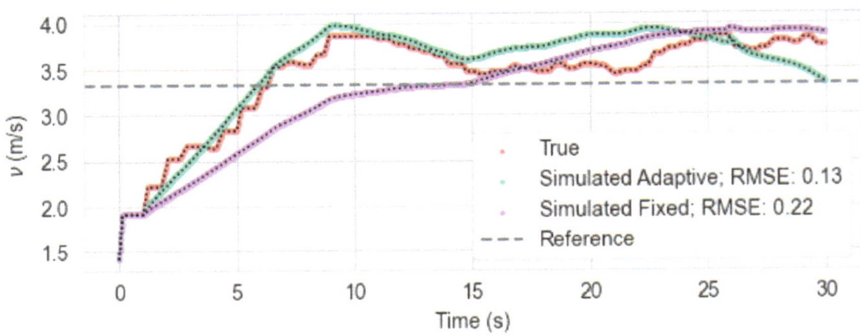

(b) 35kg rider

Fig. 7 Fixed and parametric model comparison. This figure compares the open-loop simulation of vehicle dynamics between fixed and parametric models for two trials in the e-bike dataset

parameters. Also, large-scale validation is required to ensure robustness. To this end, we are promoting the creation of a shared dataset that includes diverse roads, rider behaviors, and urban environments from multiple cities. This will enable better model comparison and validation across different studies.

References

Brankovic A, Berretta D, Formentin S, Corno M, Savaresi SM (2015) Modeling and speed limitation control of an electric kick scooter. In: 2015 European control conference (ECC). IEEE, pp 697–702. https://doi.org/10.1109/ECC.2015.7330623

Cafiso S, Di Graziano A, Marchetta V, Pappalardo G (2022) Urban road pavements monitoring and assessment using bike and e-scooter as probe vehicles. Case Stud Construct Mater 16:1–18. https://doi.org/10.1016/j.cscm.2022.e00889

Campisi T, Akgün N, Tesoriere G (2020) An ordered logit model for predicting the willingness of renting micro mobility in urban shared streets: a case study in Palermo, Italy. In: International conference on computational science and its applications. Springer, pp 796–808. https://doi.org/10.1007/978-3-030-58802-1_57

Christoforou Z, de Bortoli A, Gioldasis C, Seidowsky R (2021) Who is using e-scooters and how? Evidence from Paris. Transp Res Part D: Transp Environ 92:1–15. https://doi.org/10.1016/j.trd.2021.102708

Corno M, Giani P, Tanelli M, Matteo Savaresi S (2014) Human-in-the-loop bicycle control via active heart rate regulation. IEEE Trans Control Syst Technol 23(3):1029–1040. https://doi.org/10.1109/TCST.2014.2360912

Corno M, Roselli F, Savaresi S (2016) Bilateral control of SeNZA–A series hybrid electric bicycle. IEEE Trans Control Syst Technol 25(3):864–874. https://doi.org/10.1109/TCST.2016.2587243

De Bortoli A, Christoforou Z (2020) Consequential LCA for territorial and multimodal transportation policies: method and application to the free-floating e-scooter disruption in Paris. J Clean Prod 273:1–15. https://doi.org/10.1016/j.jclepro.2020.122898

Dhole H, Sutaone M, Vyas V (2019) Anomaly detection using convolutional spatiotemporal autoencoder. In: 2019 10th International conference on computing, communication and networking technologies (ICCCNT). IEEE, pp 1–5. https://doi.org/10.1109/ICCCNT45670.2019.8944523

Di Graziano A, Marchetta V, Cafiso S (2020) Structural health monitoring of asphalt pavements using smart sensor networks: a comprehensive review. J Traffic Transp Eng (English Edn) 7(5):639–651. https://doi.org/10.1016/j.jtte.2020.08.001

European Commission (2020) Stepping up Europe's 2030 climate ambition investing in a climate-neutral future for the benefit of our people. J Chem Inf Model 53(9):1689–1699

Ewert A, Brost M, Eisenmann C, Stieler S (2020) Small and light electric vehicles: An analysis of feasible transport impacts and opportunities for improved urban land use. Sustainability 12(19):1–17. https://doi.org/10.3390/su12198098

Felipe J, Carlos Amarillo J, Eugenio Naranjo J, Serradilla F, Díaz A (2015) Energy consumption estimation in electric vehicles considering driving style. In: 2015 IEEE 18th international conference on intelligent transportation systems, pp 101–106. https://doi.org/10.1109/ITSC.2015.25

Gelmini S, Strada SC, Tanelli M, Savaresi SM, Biase V (2020) Online assessment of driving riskiness via smartphone-based inertial measurements. IEEE Trans Intell Transp Syst (2020) https://doi.org/10.1109/TITS.2020.2987877

Gorges C, Öztürk K, Liebich R (2018) Road classification for two-wheeled vehicles. Veh Syst Dyn 56(8):1289–1314. https://doi.org/10.1080/00423114.2017.1413197

Haas T, Sander H (2020) Decarbonizing transport in the European union: emission performance standards and the perspectives for a European green deal. Sustainability 12(20):1–15. https://doi.org/10.3390/su12208381

Hassen DB, Miladi M, Slim Abbes M, Caglar Baslamisli S, Chaari F, Haddar M (2019) Road profile estimation using the dynamic responses of the full vehicle model. Appl Acoust 147:87–99. https://doi.org/10.1016/j.apacoust.2017.12.007

Hu F (2015) Road profile recovery using vertical acceleration data

Kejun J, Chenmin H, Songli Z, Pianpian X, Sha C (2021) Transport scenarios for China and the role of electric vehicles under global 2 °C/1.5 °C targets. Energy Econ 103:1–8. https://doi.org/10.1016/j.eneco.2021.105172

Knowles M, Scott H, Baglee D (2012) The effect of driving style on electric vehicle performance, economy and perception. Int J Electr Hybrid Veh 4(3):228–247. https://doi.org/10.1504/IJEHV.2012.050492

Lefever J, Berckmans D, Aerts J-M (2014) Time-variant modelling of heart rate responses to exercise intensity during road cycling. Eur J Sport Sci 14:406–412. https://doi.org/10.1080/17461391.2012.708791

Leoni J, Lucchini A, Carla Strada S, Tanelli M, Matteo Savaresi S (2023) Active driver assistance systems for e-scooters based on road quality and driving style estimation. IFAC-PapersOnLine 56(2):1977–1982. https://doi.org/10.1016/j.ifacol.2023.10.1091

Leoni J, Tanelli M, Carla Strada S, Savaresi S (2024) Safety in e-scooters: a machine-learning approach for online second passenger detection. IFAC-PapersOnLine 58(10):14–21. https://doi.org/10.1016/j.ifacol.2024.07.312

Leoni J, Vedani D, Tanelli M (2025) Adaptable, rider-in-the-loop and road-quality-aware model for e-bike dynamics. TBD, TBD(TBD):TBD

Ma Q, Yang H, Mayhue A, Sun Y, Huang Z, Ma Y (2021) E-scooter safety: the riding risk analysis based on mobile sensing data. Accident Anal Prevent 1–12:105954. https://doi.org/10.1016/j.aap.2020.105954

Masson-Delmotte V, Zhai P, Pörtner H-O, Roberts D, Skea J, Shukla PR, et al (2022) Global Warming of 1.5 C: IPCC special report on impacts of global warming of 1.5 C above pre-industrial levels in context of strengthening response to climate change, sustainable development, and efforts to eradicate poverty. Cambridge University Press. https://doi.org/10.1017/9781009157940

Meyer D, Zhang W, Tomizuka M (2015) Sliding mode control for heart rate regulation of electric bicycle riders. In: Dynamic systems and control conference. American Society of Mechanical Engineers, pp 1–10. https://doi.org/10.1115/DSCC2015-9712

Michelangelo-Gulino S, Vichi G, Zonfrillo G, Vangi D (2022) Comfort assessment for electric kick scooter decks. In: IOP conference series: materials science and engineering, vol 1214. IOP Publishing, pp 1–12. https://doi.org/10.1088/1757-899X/1214/1/012043

Nguyen T, Lechner B, Diew Wong Y (2019) Response-based methods to measure road surface irregularity: a state-of-the-art review. Eur Transp Res Rev 11:1–18. https://doi.org/10.1186/s12544-019-0380-6

Radrizzani S, Panzani G, Savaresi SM (2023) Virtual-bike emulation in a series-parallel human-powered electric bike. IFAC-PapersOnLine 56(2):2558–2563. https://doi.org/10.1016/j.ifacol.2023.10.1338

Rath JJ, Chakravarthy Veluvolu K, Defoort M (2014) Simultaneous estimation of road profile and tire road friction for automotive vehicle. IEEE Trans Veh Technol 64(10):4461–4471. https://doi.org/10.1109/TVT.2014.2373434

Sarzynski MA, Rankinen T, Earnest CP, Leon AS, Rao DC, Skinner JS, Bouchard C (2013) Measured maximal heart rates compared to commonly used age-based prediction equations in the heritage family study. Am J Hum Biol 25(5):695–701. https://doi.org/10.9790/6737-0310408

Sayers MW, Gillespie TD, Queiroz CAV (1986) The international road roughness experiment: Establishing correlation and a calibration standard for measurements. Technical report. University of Michigan, Ann Arbor, Transportation Research Institute

Sopjani L, Janhager Stier J, Ritzén S, Hesselgren M, Georén P (2019) Involving users and user roles in the transition to sustainable mobility systems: the case of light electric vehicle sharing in Sweden. Transp Res Part D: Transp Environ 71:207–221. https://doi.org/10.1016/j.trd.2018.12.011

Su SW, Wang L, Celler BG, Savkin AV, Guo Y (2007) Identification and control for heart rate regulation during treadmill exercise. IEEE Trans Biomed Eng 54(7):1238–1246. https://doi.org/10.1109/TBME.2007.890738

Tabatabaie M, He S (2023) Naturalistic e-scooter maneuver recognition with federated contrastive rider interaction learning. In: Proceedings of the ACM on interactive mobile wearable and ubiquitous technologies, vol 6, issue 4, pp 1–27. https://doi.org/10.1145/3570345

Teixeira JF, Diogo V, Bernát A, Lukasiewicz A, Vaiciukynaite E, Stefania Sanna V (2023) Barriers to bike and e-scooter sharing usage: an analysis of non-users from five European capital cities. Case stud Transp Policy 13:1–14. https://doi.org/10.1016/j.cstp.2023.101045

Van Bree B, Verbong GPJ, Kramer GJ (2010) A multi-level perspective on the introduction of hydrogen and battery-electric vehicles. Technol Forecast Soc Change 77(4):529–540. https://doi.org/10.1016/j.techfore.2009.12.005

Wan N, Alireza Fayazi S, Saeidi H, Vahidi A (2014) Optimal power management of an electric bicycle based on terrain preview and considering human fatigue dynamics. In 2014 American control conference, pp 3462–3467 https://doi.org/10.1109/ACC.2014.6859373

Yeganeh SF, Mahmoudzadeh A, Amin Azizpour M, Golroo A (2019) Validation of smartphone based pavement roughness measures. arXiv:1902.10699, https://doi.org/10.48550/arXiv.1902.10699

Yousefzadeh M, Azadi S, Soltani A (2010) Road profile estimation using neural network algorithm. J Mech Sci Technol 24:743–754. https://doi.org/10.1007/s12206-010-0113-1

A Systematic Approach to Enhance Road Safety of Electric Pedal-Assisted Vehicles for Future Urban Mobility

Andrea Raimondi, Giovanni Savino, Niccolò Baldanzini, Jonathan Lagrimino, Simona Viciani, Giulia Biagioni, Jessica Leoni, Mirko Mazzoleni, Davide Previtali, Luigi Gaioni, and Marcello Minervini

Abstract At a global level, the World Health Organization (WHO) estimated 1.19 million road traffic deaths in 2021; among these, cyclists account for 5% of fatalities. In that regard, a systematic and general approach to road safety through data analysis was proposed in the 1970s by the first administrator of the National Highway Traffic Safety Administration (NHTSA), Dr. William Haddon, who analyzed the relationship

---

A. Raimondi (✉) · G. Savino · N. Baldanzini
Department of Industrial Engineering (DIEF), University of Florence, Florence, Italy
e-mail: andrea.raimondi@unifi.it

G. Savino
e-mail: giovanni.savino@unifi.it

N. Baldanzini
e-mail: niccolo.baldanzini@unifi.it

J. Lagrimino
Department of Architecture (DIDA), University of Florence, Florence, Italy
e-mail: jonathan.lagrimino@unifi.it

S. Viciani · G. Biagioni
Department of Legal Sciences (DSG), University of Florence, Florence, Italy
e-mail: simona.viciani@unifi.it

G. Biagioni
e-mail: giulia.biagioni@unifi.it

J. Leoni
Department of Electronics, Information and Bioengineering (DEIB), Polytechnic University of Milan, Milan, Italy
e-mail: jessica.leoni@polimi.it

M. Mazzoleni · D. Previtali
Department of Management, Information, and Production Engineering (DIGIP), University of Bergamo, Dalmine, Italy
e-mail: mirko.mazzoleni@unibg.it

D. Previtali
e-mail: davide.previtali@unibg.it

L. Gaioni · M. Minervini
Department of Engineering and Applied Sciences (DISA), University of Bergamo, Dalmine, Italy

© The Author(s), under exclusive license to Springer Nature Switzerland AG 2026
R. Moscarelli et al. (eds.), *The Future of Slow Mobility*, Research for Development,
https://doi.org/10.1007/978-3-032-23660-9_9

between the phases of a road crash and traffic factors to define the well-established "Haddon Matrix". This approach is still of great relevance today. The study described in this book chapter extends Dr. Haddon's methodology by engaging experts from different research fields. The uniqueness of this approach lies in its focus on technological, socio-cultural, and economic countermeasures for light pedal vehicles, such as e-bikes and e-cargo bikes, which to the best of authors' knowledge, has never been addressed in previous analyses. This modified version of the Haddon Matrix is an asset tailored to be more effective than the original to address the challenges of the European mobility context.

1 Introduction

The Global Status report on road safety published by the World Health Organization (WHO) in 2023 indicated a 5% decrease in global road traffic fatalities between 2010 and 2021 (from 1.25 to 1.19 million), despite increases in population and vehicle fleet (World Health Organization 2024). Road crashes remain the leading cause of death for individuals aged 5 to 29 and the 12th leading cause of death globally. Most victims are motorcyclists (30%), car occupants (25%), pedestrians (21%), and cyclists (5%). Also, efforts to enhance road safety, such as legislation addressing speed limits, drink-driving, helmet use, and seat belts, have made limited progress, with only seven countries meeting WHO best practice standards.

The "Safe System" (SS) approach, prioritizing people and safety, has proven to be more effective in Europe and the Western Pacific, demonstrating achievable targets with sufficient political commitment (U.S. Department of Transportation 2025).

Italian road trauma statistics also raised concerns for 'Vulnerable Road Users' (VRUs), including light pedal vehicle users. For instance, in 2023, cyclists' fatalities reached 212, a 3.4% increase from 2022 (Istat 2024). This underscores that promoting light pedal vehicles for sustainable urban transport necessitates mandatory safety initiatives, as increased VRUs use, without preventive countermeasures, escalates crash risk. Accordingly, this book chapter explores the application of the well-established 'Haddon Matrix' (HM) as a systematic framework for improving cycling safety by addressing risks across the different crash phases (Haddon 1972). The objective of this study is to identify safety-related areas, treatments, interventions, and countermeasures, for a safe design, implementation, and use of light pedal vehicles, particularly e-bikes and e-cargo bikes.

To address current mobility challenges, multidisciplinary experts were involved in the study to provide advice addressing this objective. This led to the definition of

e-mail: luigi.gaioni@unibg.it

M. Minervini
e-mail: marcello.minervini@unibg.it

an adapted version of the HM, tailored to this scenario, including relevant items and observations to enhance road safety of light pedal vehicles.

2 Background

Recent advancements in road safety research reject serious injuries and fatalities as inevitable outcomes of road transport, trying to define and promote new guiding principles such as ‘Life is not negotiable’ from Vision Zero, an initiative launched in Sweden in the 1990s (Vision Zero Network 2025).

Vision Zero was influential for road safety policies worldwide, leading to the commitment of governments towards the zero-fatality goal in Europe, USA, and Australia, to name a few. In that regard, the European Commission confirmed this goal for 2050 with the “Sustainable and Smart Mobility Strategy” (European Commission 2020).

Institutions committed to Vision Zero have started adopting the SS approach. An example is represented by the US National Highway Traffic Safety Administration (NHTSA) (U.S. Department of Transportation 2025).

Since the 1970s, NHTSA has used a systematic, data-driven approach to investigate road safety challenges and identify proper countermeasures.

Within this context, Dr. Willian Haddon, NHTSA’s first administrator, proposed a simple but brilliant method that professionals can use for analyzing all the relevant aspects of the traffic system that have a role in road crash losses (Haddon 1972). The method considers the three phases of a road crash (pre-crash, crash, and post-crash) and their relationship with the three main traffic factors (user, vehicle, and environment). Losses due to energy-damage are produced throughout a sequence of three phases in which the relevant factors intervene and interact; the matching of the three phases with the three factors that influence the crash occurrence constitutes the so-called “Haddon Matrix”. Each HM cell features a list of keywords related to one specific crash phase and influenced by one specific factor. Each keyword in the table warrants an analysis to understand the phenomenon and identify possible countermeasures. For example, when focusing on the human factor (user) and the pre-crash phase, one possible element is crash avoidance maneuvers, such as braking. The analysis of real-world crash data determines the influence that braking has on crash occurrence and the related damage (e.g., serious injuries). Once the associated factors and the causative factors are identified, possible countermeasures can be determined to reduce crash occurrence or mitigate the related injury outcomes.

Overall, the HM provides a useful framework for systematic road safety analysis, being aligned with the SS. Among the applications of the HM, recent examples are available in the literature (Wallington et al. 2014; Batala Mpondo et al. 2014; Ncube et al. 2016; Baru et al. 2019; Ewbank et al. 2019; Sadeghi-Bazargani et al. 2020; Sundet et al. 2020; Zou et al. 2020; Das et al. 2024).

3 Methodology Employed

Building upon the original structure of the HM, an enhanced version, more tailored to the needs of the actual mobility system, is here defined.

Several experts from different research fields were consulted during the study, contributing to the definition of an adapted HM, filled with pre-existing or innovative safety-related items, along with a list of relevant observations.

Naturally, the presented version - illustrated in Sect. 3.2—should be regarded as preliminary. Accordingly, it has been intentionally designed to support collaborative development, serving as a foundation for fostering connections among the experts, facilitating the further extension and refinement of its content, as described in Sect. 4.3.

3.1 Basic Haddon Matrix

The original layout of the HM has three columns—the factors of the epidemiological phenomenon—and three rows—the temporal phases of the crash event—as depicted in Fig. 1. The factors are simply identified as follows:

- Human (human users of the road infrastructure)
- Vehicle and equipment (vehicles using the road infrastructure)

Fig. 1 The basic Haddon Matrix

	Human	*Vehicle*	*Environment*
Pre-crash			
Crash			
Post-crash			

- Environment (road infrastructure).

The phases are categorized as:

- Pre-crash (before the crash event, while using the vehicle)
- Crash (while the energy exchange occurs)
- Post-crash (after the crash event, both at the scene and post-event).

3.2 Enhanced Haddon Matrix

Although the basic HM highlighted several key aspects of road safety, its application to this specific context revealed gaps that required the inclusion of additional elements. To meet the needs of the target scenario, the factors and phases were further refined, leading to an enhanced HM, illustrated in Fig. 2. This extension was primarily motivated by the broader perception of road safety issues and the necessity for increased granularity across various domains.

In our study, the proposed list of factors is the following:

- Human (road users)

	Human	*Vehicle and equipment: physical characteristics*	*Vehicle and equipment: movement and location*	*Environment: physical characteristics*	*Environment: socio-cultural and economic characteristics*	*Management culture and leadership*	*Journey management*
Pre-drive							
Drive and Pre-crash							
Crash							
Post-crash							
Post-event							

Fig. 2 The enhanced Haddon Matrix

- Vehicle and equipment: physical characteristics (vehicle geometry or materials, type of installed systems and components, etc.)
- Vehicle and equipment: movement and location (vehicle status)
- Environment: physical characteristics (road layout and geometry, type of road surface, etc.)
- Environment: social-cultural-economic characteristics (community attitudes towards road safety)
- Management culture and leadership (society as it can become through high-level initiatives)
- Journey management (planning actions for mobility operations).

The phases were refined from the three basic elements to the following:

- Pre-drive (before getting to the vehicle)
- Drive and Pre-crash (before the crash event, while using the vehicle)
- Crash (while the energy exchange occurs)
- Post-crash (at scene, after a crash occurs)
- Post-event (following a crash, not at the crash scene).

3.3 *Experts' Contribution*

As previously mentioned, this study was carried out with input from experts across several research fields, gathered through an initial survey. The survey was structured in sections, as described below:

- Section 1: introduction to the study and its scope
- Section 2: request for the experts' professional details (professional role, scientific background, etc.)
- Section 3: explanation of the enhanced HM, presented with selectable cells representing the safety-related areas, and request for the selection of each expert's areas of interest. An existing subsection allowed the experts to provide details for the selected areas (e.g., the interest in one or more keywords already present in the HM).

The experts were individually contacted for a telematic meeting, providing a further explanation of the HM and the aim of the study. They were invited to confirm or modify the selected area of interest, and to correlate each area with one or more keywords present in the enhanced HM, or to identify them as a further extension.

In addition, the experts were requested to write one or more short paragraphs to facilitate focused insights on each safety-related area, structured as follows:

- Research output (to be identified within the study)
- Inputs required: information to achieve the desired output
- Theoretical framework: reference to a consolidated knowledge within the research fields (e.g., a well-known behavioral model)
- Open questions.

It was clarified that each paragraph had to focus on a distinct keyword related to a specific area of concern.

Technical support was provided for the proper alignment of the expertise of each expert with the identified keywords.

4 Results

Eleven participants contributed to the study, comprising the authors and co-authors of this book chapter. All of them have an academic background, ranging from early-stage to advanced careers, as summarized in Table 1. Overall, three main professional profiles were identified: engineer, designer, and lawyer.

The established multidisciplinary team brought together a wealth of expertise, ranging from public health to engineering and social sciences, ensuring a comprehensive approach to the objective of the study. Their expertise span road safety, including road crash investigation and active safety technologies for Power Two Wheelers (PTW), and vehicle technology, encompassing powertrain systems design, smart sensors, Internet of Things (IoT), Battery Management Systems (BMS) algorithms for e-bikes, and battery modelling. Additionally, their proficiency extends to the design and regulatory framework of light pedal vehicles, with a primary focus on emerging frontiers of civil liability in light mobility.

Table 1 Experts' professional profiles

Expert number	Professional role	Scientific background	Years of working experience
1	Associate professor	Civil law	29
2	Associate professor	Mechanical engineering	25
3	Associate professor	Computer engineering	19
4	Associate professor	Mechanical engineering	18
5	Researcher	Computer engineering	11
6	Researcher	Biomedical engineering	6
7	Researcher	Electrical engineering	6
8	Researcher	Computer engineering	6
9	Ph.D. Student	Civil law	4
10	Ph.D. Student	Mechanical engineering	2
11	Ph.D. Student	Product design	2

4.1 Experts' Contributions: Adapted Haddon Matrix

Each expert's contribution was reviewed, verifying the alignment of the identified keywords with each safety-related area. An adapted HM was then obtained, shown in Fig. 3. The green cells contain the keywords identified by the experts as already available in the enhanced HM, while the yellow cells contain the newly identified keywords. The white cells contain the keywords included in the enhanced HM but not selected or identified by the experts.

Compared to the enhanced HM, Fig. 4 illustrates the adapted HM with color gradients that reflect the volume of contributions received for each safety-related area, while also highlighting the scientific background of the experts who provided inputs among the different areas.

Engineers and designers focused on the physical characteristics of light pedal vehicles for the first three phases, while lawyers prioritized human factors in boundary phases and the socio-cultural and economic environmental characteristics for the pre-crash phase.

Although several contributions were received both for the new phases and factors of the enhanced HM, the covered areas were mainly related to the factors and phases of the basic HM. Still, new keywords were defined, allowing for a detailed analysis and description of the related initiatives proposed to enhance cycling safety, which will be presented in the next Section.

4.2 Pre-drive Phase: Prevention Through Strategic Readiness

The pre-drive phase focuses on proactive measures for cyclists and their vehicles to ensure safe journeys.

Improving human factors plays a pivotal role in this phase. In that regard, comprehensive safety education programs tailored to the diverse needs of cyclists are reported as crucial; for instance, teenagers using bicycles for leisure, commuters riding within urban traffic, and non-residents unfamiliar with local road signs all require distinct educational approaches. To this end, a particularly promising initiative is the introduction of a voluntary safety license, which could be endorsed by public institutions, delivery platforms, or private organizations, encouraging participants to engage in structured safety training, thereby promoting a culture of responsible cycling. Nevertheless, there is a limited number of studies in the literature addressing this topic from a normative perspective (Magaña et al. 2021; Laverdet et al. 2023; Niemann et al. 2023).

Vehicle and equipment factors' readiness is also important. Predictive maintenance of systems offers a proactive approach to enhance the reliability of bicycles, particularly e-bikes. In fact, real-time diagnostics can detect issues such as motor faults or wear, preventing sudden breakdowns that could compromise cycling safety.

	Human	*Vehicle and equipment: physical characteristics*	*Vehicle and equipment: movement and location*	*Environment: physical characteristics*	*Environment: socio-cultural and economic characteristics*	*Management culture and leadership*	*Journey management*
Pre-drive	Induction process (safety education)	Vehicle selection	Speed management	Road design	Speed limits (survivability)	Online data-warehouse and tools for decision making	Reducing need to travel
	Handbook	Road worthiness of vehicles		Route signage	Policing and law enforcement (e.g., alcohol use)	Risk analysis	Journey planning
	Training needs	Maintenance			Engagement with research community	Pilot new solutions	Route selection
	Targeted coaching based on risks	Vehicle utilisation			Alcohol/drug policy		
	Encouragement	Predictive maintenance					
	Proactive programmes	Components design					
	Fitness to ride						
	Voluntary driving license training course						
	Safety guidelines and protocols						
Drive & Pre-crash	Behaviour	Handling	Path choice	Road layout	Speeding behaviour	Reporting process	Fatigue management
	Impairment	Brakes	Speed	Pedestrian facilities	Red light running		Re-routing
	User conspicuity	Vehicle conspicuity	Traffic interaction	Signs	Pay How You Drive (PHYD) and Pay As You Drive (PAYD) index-based insurances		
	Use of Personal Protective Equipment (PPE)	Intelligent Transportation System (ITS)	Telemetry (to manage and monitor use)	Low visibility conditions (e.g., absence of streetlights)			
	Monitoring (e.g., for distraction or fatigue)	Mechanical failures					
	Fitness to ride (e.g., to properly react)	Active safety technologies					
	Signal processing for monitoring or detecting user actions	Regenerative braking (e.g., to avoid critical State of Charge)					
		Dynamic vehicle control					
		Electronic controller (e.g., for a safe transition from electric to human traction)					
Crash	PPE effectiveness	Crash-protective design (crashworthiness)	Telemetry (to capture data)	Crash-protective roadside object	Escalation process	Reporting process and support to drivers	Manage scene
Post-crash	Physiological monitoring	Automatic crash notification (E-call)	Global Navigation Satellite System (GNSS)	Rescue facilities	Rescue procedures	Support to drivers	
	Clear process to manage scene and report events			Road congestion	First aid skills		
					Emergency medical services		
Post-event	Driver debrief	Vehicle inspection and repair	Telemetry data investigation		Community learning process management	Event investigation	Debrief and review
	Counselling, support and rehabilitation					Program renewal	
	Legal assistance (e.g., to support mental well-being or effective medical treatment)						

Fig. 3 The adapted Haddon Matrix. The green and white cells contain the keywords presented in the enhanced HM, with the green cells representing those selected by the experts. The yellow cells highlight the newly identified keywords

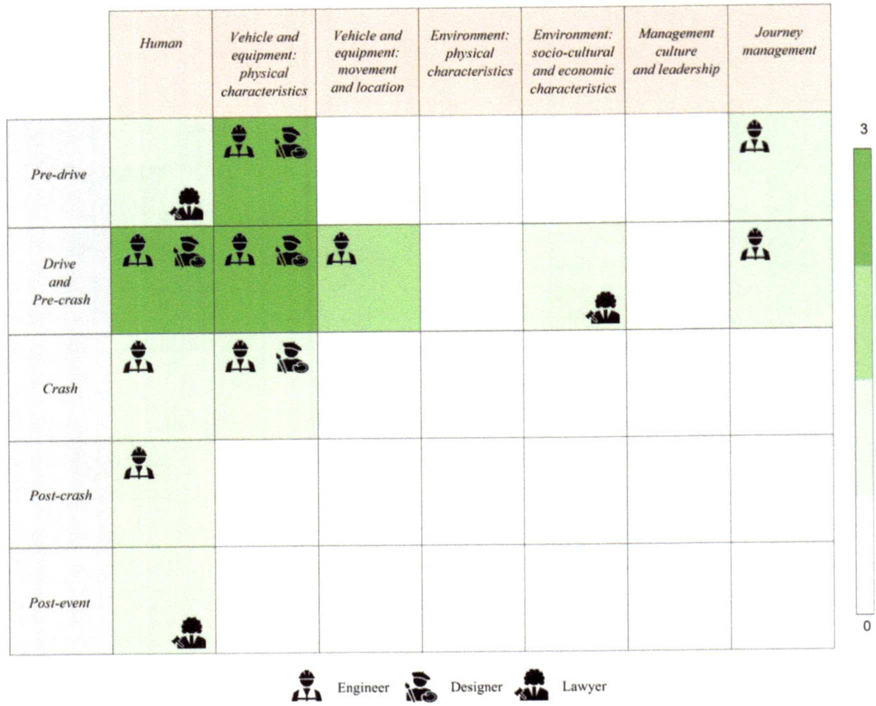

Fig. 4 The adapted Haddon Matrix builds upon its enhanced version, showing the density of the experts' contributions for each safety-related area, and their scientific backgrounds

Additionally, the analysis of historical data enables the identification of latent defects, supporting proper maintenance strategies (Lee et al. 2020).

Ultimately, journey management factors need consideration. By combining data on the battery State of Charge (SoC), rider behavior, and road characteristics (e.g., type of road, or pavement conditions), it is possible to provide accurate range predictions, reducing the likelihood of a cyclist being stuck on the road. This is particularly important in high-risk areas, such as zones with heavy traffic and lack of dedicated bicycle lanes (Burani et al. 2022). Consequently, cyclists can select or plan safer routes based on their preferences, even enhancing the comfort of cycling trips.

4.3 Drive and Pre-crash Phase: Mitigating Road Hazards

During the drive and pre-crash phase, the focus changes, as cyclists are now riding within real-world traffic contexts.

The "Fitness to ride" is a critical element, as cyclists' physical readiness directly impacts on their ability to manage unexpected events. Wearable devices can help monitor this phenomenon, as they allow for tracking vital signs, thus assessing cyclist's physiological condition in real-time (Boonnithi and Phongsuphap 2011). Accordingly, (Niewiadomski et al. 2007; Santos et al. 2020) propose an approach to prevent excessive fatigue and checks that cyclists remain alert and capable to effectively respond to potential hazards, even suggesting rest breaks when fatigue or distraction levels become critical. However, it is important to note that eliminating fatigue is neither feasible nor desirable, as it would undermine the health benefits of physical exercise. Hence, this measure is primarily intended for untrained or older cyclists, who benefit most from guidance in managing fatigue while riding.

Active safety technologies further enhance cycling safety by intervening on vehicle dynamics or providing clear warnings about road hazards, allowing cyclists to promptly maintain vehicle control and avoid potential collisions (Kapousizis et al. 2023). These technologies can enable e-bikes to adapt to varying road conditions and dynamically respond to changes, ensuring optimal handling and preventing crashes in scenarios which require sudden maneuvers (Corno et al. 2017, 2022; Guanetti et al. 2017). Similarly, speed regulation technologies, which integrate Global Positioning System (GPS) data and real-time sensor inputs, ensure that e-bikes operate within safe limits tailored to the environment and rider condition (Brankovic et al. 2015). Moreover, smooth transitions between pedal power only and pedal-assisted modes further enhance rider confidence, especially during risk scenarios and when the battery SoC is critically low. Despite the promising results achieved in the automotive sector, where these technologies are being increasingly adapted, they are still in the early design phase for e-bikes, thus requiring further research efforts (Kapousizis et al. 2023).

Ultimately, it must be noted that also socio-cultural and economic approaches can contribute to mitigating road hazards. Behavioral insurance policies, such as Pay How You Drive (PHYD) and Pay As You Drive (PAYD), promote responsible cycling behaviors by linking premiums to factors such as speed regulation, braking patterns, and adherence to traffic laws (Yang et al. 2022). These schemes not only encourage safer practices but also make insurance more affordable, thereby improving its accessibility and appeal to a broader range of cyclists.

4.4 Crash Phase: Improving Vehicle Resilience and User Protection

The crash phase is mainly centered on vehicle crashworthiness and user protection.

To enhance safety, road vehicles are increasingly designed with collapsible structures, often using different materials to better withstand impacts and mitigate injury severity of VRUs (Carter and Neal-Sturgess 2009; McNally and Rosenberg 2013; Sun et al. 2018; Ptak et al. 2020). In that regard, advanced multibody simulation tools play a prominent role to support researchers and manufacturers to analyze the dynamics of different crash scenarios and optimize vehicle design to prioritize cyclist protection (Siemens 2025).

Moreover, research also focus on Personal Protective Equipment (PPE), including protective clothing (Pallacci et al. 2019) and innovative helmets equipped with airbags (Mathon et al. 2023), improving their design to further reduce injury severity by effectively absorbing impact forces directly transmitted to users.

Besides minimizing the physical consequences of crashes, these solutions also improve overall confidence among users in choosing light pedal vehicles as a mode of transport.

4.5 Post-crash Phase: Coordinated Emergency Response and Recovery Efforts

The post-crash phase focuses on safety management in the immediate aftermath of a crash to reduce harm and ensure prompt interventions.

Physiological monitoring devices are instrumental in this context. For instance, intelligent PPEs as smart helmets, which are capable of correlating measured head accelerations with head injury levels (Dharani et al. 2020), and wearable devices that measure vital signs, can transmit real-time data of the physiological status of cyclists (Boonnithi and Phongsuphap 2011; Niewiadomski et al. 2007; Santos et al. 2020). This information, if promptly provided to emergency responders, enables more effective medical interventions, thereby improving the chances of cyclist survival, especially in remote areas.

Legal and financial recovery mechanisms are also important. Specifically, insurance policies offering legal assistance can streamline the complex and stressful process of filing claims and settling disputes. By ensuring access to affordable and efficient legal support, these policies can alleviate the psychological and financial burden on cyclists, enabling them to focus on recovery.

4.6 Post-event Phase: Long-Term Improvements

The post-event phase extends beyond immediate recovery, focusing on extracting lessons from crashes to inform future safety interventions.

Hence, data collection and analysis are central to this phase (Yang and Yang 2023). By anonymizing and aggregating information from crashes, such as braking behavior, road conditions, and environmental factors, researchers and policymakers can identify patterns and underlying causes. Moreover, this allows the evaluation and refinement of on-board systems to ensure continuous technological improvement.

This phase also highlights significant ethical concerns, with particular emphasis on data privacy. Establishing robust safeguards and transparent usage policies is essential to ensure public trust and compliance with regulatory frameworks.

4.7 Relevant Items and Observations

Several observations emerged from the study, allowing to define a list of key interventions to enhance cycling safety.

In that regard, it must be noted that some elements presented in the adapted HM were already present in the enhanced HM, thus confirming their importance for road safety, and also underlining the inherent multidisciplinarity among the experts. An example is "Monitoring", related to cyclists' distraction or fatigue, which may be monitored using on-board and wearable devices.

Additionally, a comparable number of elements were newly identified, thereby proving the potential of the HM and its versatility. In more detail, the involvement of legal experts highlighted two main elements relevant to light mobility: insurance policies and legal assistance. Furthermore, other new elements provided interesting insights across various research fields, such as "Re-routing", associated with the dynamic vehicle control to modify a route in real-time, accounting for the safety-related factors that characterized a trip.

Ultimately, while a few safety-related areas remained partially or completely uncovered, these should not be disregarded, as innovative items relevant to enhancing cycling safety could still emerge.

5 Conclusions

This study demonstrates the potential of the Haddon Matrix as a systematic approach for addressing road safety of light pedal vehicles, as those under development within the MOST—Spoke 5, identifying effective countermeasures to support their design and implementation.

The adopted methodology proved to be effective and versatile, both in its ability to establish connections within a multidisciplinary research context and to gather useful information for identifying specific and relevant elements within the safety-related areas of the HM. In that regard, some original observations were provided; their importance was confirmed, as they aligned with already-known concepts or emerged as innovative.

However, it must be noted that several competencies were excluded, although they may be influential for addressing safety of light pedal vehicles.

Therefore, future research will involve the experts' re-engagement to assess whether any elements present the adapted HM, or newly identified ones align with ongoing research activities or projects, providing a basis for revisiting the HM content in light of emerging methodologies or technologies aimed at enhancing cycling safety.

Overall, the insights of this study underscore the potential of the HM for understanding and mitigating crash risks, laying the groundwork for long-term improvements in European mobility contexts.

Acknowledgements This work is funded by the European Union—Next Generation EU. "Sustainable Mobility Center (Centro Nazionale per la Mobilità Sostenibile - CNMS)", M4C2—Investment 1.4, Project Code CN_00000023, CUP B13C22001000001.

References

Baru A, Azazh A, Beza L (2019) Injury severity levels and associated factors among road traffic collision victims referred to emergency departments of selected public hospitals in Addis Ababa, Ethiopia: the study based on the Haddon matrix. BMC Emerg Med 19:2. https://doi.org/10.1186/s12873-018-0206-1

Batala Mpondo G, Bouanga M, Saya YMC et al (2014) Étude exploratoire de la sécurité routière à Brazzaville et à Pointe-Noire en République du Congo. Santé Publique 26:71–79. https://doi.org/10.3917/spub.140.0071

Boonnithi S, Phongsuphap S (2011) Comparison of heart rate variability measures for mental stress detection. In: Paper presented at the computing in cardiology, Hangzou, China, 18–21 Sept 2011

Brankovic A, Berretta D, Formentin S, et al (2015) Modeling and speed limitation control of an electric kick scooter. In: Paper presented at the European control conference (ECC), Linz, Austria, 15–17 Jul 2015

Burani E, Cabri G, Leoncini M (2022) An algorithm to predict e-bike power consumption based on planned routes. Electronics 11:1105. https://doi.org/10.3390/electronics11071105

Carter EL, Neal-Sturgess CE (2009) MADYMO reconstruction of a real-world collision between a vehicle and cyclist. Int J Crashworthiness 14:379–390. https://doi.org/10.1080/13588260902823999

Corno M, Duz A, Savaresi SM (2022) Design of a charge-sustaining energy management system for a free-floating electric shared bicycle. IEEE Trans Control Syst Technol 30:654–666. https://doi.org/10.1109/TCST.2021.3071197

Corno M, Roselli F, Savaresi SM (2017) Bilateral control of SeNZA—a series hybrid electric bicycle. IEEE Trans Control Syst Technol 25:864–874. https://doi.org/10.1109/TCST.2016.2587243

Das S, Hossain A, Rahman MA et al (2024) Case study on the traffic collision patterns of e-scooter riders. Transp Res Rec 2678:575–589. https://doi.org/10.1177/03611981231185770

Dharani P, Ganesh T, Gopinath V, Vali YS (2020) Smart safety helmet for bike riders using IoT. Int Res J Multidiscip Tech 2:21–30. https://doi.org/10.34256/irjmt2044

European Commission (2020) Sustainable and smart mobility strategy-putting European transport on track for the future. In: Preparatory documents. European Union law. Available via: EUR-Lex. https://eur-lex.europa.eu/resource.html?uri=cellar:5e601657-3b06-11eb-b27b-01aa75ed71a1.0001.02/DOC_1&format=PDF. Accessed 6 Jun 2025

Ewbank C, Gupta S, Stewart BT et al (2019) A systematic review of oil tanker truck disasters: identifying prevention targets. Burns 45:905–913. https://doi.org/10.1016/j.burns.2018.12.010

Guanetti J, Formentin S, Corno M, Savaresi SM (2017) Optimal energy management in series hybrid electric bicycles. Automatica 81:96–106. https://doi.org/10.1016/j.automatica.2017.03.021

Haddon W (1972) A logical framework for categorizing highway safety phenomena and activity. J Trauma 12:193–207. https://doi.org/10.1097/00005373-197203000-00002

Istat (2024) INCIDENTI STRADALI Anno 2023. In: Comunicato stampa. Istituto Nazionale di Statistica. Available via Istat. https://www.istat.it/wp-content/uploads/2024/07/REPORT-INCIDENTI-STRADALI-2023.pdf. Accessed 6 Jun 2025

Kapousizis G, Ulak MB, Geurs K et al (2023) A review of state-of-the-art bicycle technologies affecting cycling safety: level of smartness and technology readiness. Transp Rev 43:430–452. https://doi.org/10.1080/01441647.2022.2122625

Laverdet C, Malola P, Meyer T et al (2023) Electric personal mobility device driver behaviors, their antecedents and consequences: a narrative review. J Saf Res 86:274–285. https://doi.org/10.1016/j.jsr.2023.07.006

Lee SB, Stone GC, Antonino-Daviu J et al (2020) Condition Monitoring of Industrial Electric Machines: State of the Art and Future Challenges. IEEE Ind Electron Mag 14:158–167. https://doi.org/10.1109/MIE.2020.3016138

Magaña VC, Pañeda XG, Garcia R et al (2021) Beside and behind the wheel: factors that influence driving stress and driving behavior. Sustainability 13:4775. https://doi.org/10.3390/su13094775

Mathon B, Duarte Rocha V, Py J-B et al (2023) An air-filled bicycle helmet for mitigating traumatic brain injury. Bioengineering 10:762. https://doi.org/10.3390/bioengineering10070762

McNally DS, Rosenberg NM (2013) MADYMO simulation of children in cycle accidents: a novel approach in risk assessment. Accid Anal Prev 59:469–478. https://doi.org/10.1016/j.aap.2013.07.022

Ncube PZ, Cheteni P, Sindiyandiya KP (2016) Road accidents fatalities trends and safety management in South Africa. Probl Perspect Manag 14:627–633

Niemann M, Braun KF, Otto E et al (2023) Dangers of e-mobility: a systematic review and meta-analysis of sustained injury patterns and injury severity. Saf Sci 167:106283. https://doi.org/10.1016/j.ssci.2023.106283

Niewiadomski W, Gasiorowska A, Krauss B et al (2007) Suppression of heart rate variability after supramaximal exertion. Clin Physiol Funct Imaging 27:309–319. https://doi.org/10.1111/j.1475-097X.2007.00753.x

Pallacci T, Baldanzini N, Barbani D et al (2019) Preliminary effectiveness assessment of an airbag-based device for riders' leg protection in side impacts. PSI 24:240–250. https://doi.org/10.1016/j.prostr.2020.02.021

Ptak M, Wilhelm J, Sawicki M (2020) Safety analysis of children transported on bicycle-mounted seat during traffic accidents. Int J Crashworthiness 25:612–627. https://doi.org/10.1080/13588265.2019.1626967

Sadeghi-Bazargani H, Sadeghpour A, Wilson ML et al (2020) Developing a national integrated road traffic injury registry system: a conceptual model for a multidisciplinary setting. J Multidiscip Healthc 13:983–996. https://doi.org/10.2147/JMDH.S262555

Santos P et al (2020) Real time mental stress detection through breath analysis. In: Camarinha-Matos L, Farhadi N, Lopes F, Pereira H (eds) Technological innovation for cyber-physical systems. 11th Advanced doctoral conference on computing, electrical and industrial systems, Costa de Caparica, Portugal, 2020. IFIP advances in information and communication technology, vol 577. Springer, Cham, p 345

Siemens (2025) Simcenter Madymo software. https://www.siemens.com/global/en.html. Accessed 6 Jun 2025

Sun J, Li Z, Pan S et al (2018) Identification of pre-impact conditions of a cyclist involved in a vehicle–bicycle accident using an optimized MADYMO reconstruction combined with motion capture. J Forensic Leg Med 56:99–107. https://doi.org/10.1016/j.jflm.2018.03.014

Sundet M, Mulima G, Kajombo C et al (2020) Adult pedestrian and cyclist injuries in Lilongwe, Malawi: a cross-sectional study. Malawi Med J 32:197–204

U.S. Department of Transportation (2025) What is a safe system approach? https://www.transportation.gov. Accessed 6 Jun 2025

Vision Zero Network (2025) What is vision zero? https://visionzeronetwork.org/about/what-is-vision-zero. Accessed 6 Jun 2025

Wallington D, Murray W, Darby P et al (2014) Work-related road safety: Case study of British Telecommunications (BT). J Transp Policy 32:194–202. https://doi.org/10.1016/j.tranpol.2014.01.002

World Health Organization (2024) Global status report on road safety 2023. https://www.who.int. Accessed 6 Jun 2025

Yang Z, Wei Y, Sang J et al (2022) A comprehensive review of driving style evaluation approaches and product designs applied to vehicle usage-based insurance. Sustainability 14:7705. https://doi.org/10.3390/su14137705

Yang Z, Yang Z (2023) Key contributory factors influencing cycling safety: comprehensive review based on accident data and literature survey. ASCE ASME J Risk Uncertain Eng Syst A Civ Eng 9:04023018. https://doi.org/10.1061/AJRUA6.RUENG-1014

Zou X, Vu HL, Huang H (2020) Fifty years of accident analysis & prevention: a bibliometric and scientometric overview. Accid Anal Prev 144:105568. https://doi.org/10.1016/j.aap.2020.105568

How Cities Transform When Slow and E-Mobility Shapes Urban Planning and Design

Aysegul Gungor and Nicola Petaccia

Abstract This study explores how cities can be transformed through the lens of slow mobility, encompassing active and electric modes such as walking, cycling and micromobility, made increasingly viable by the e-mobility revolution. It challenges traditional, car-centric urban design by analyzing how different mobility modes occupy space and how this space can be reclaimed and redesigned through slow mobility to improve public life. Following a methodology which is structured around two main components, it conducts a spatial occupancy analysis that compares slow and fast mobility in terms of movement and staying and develops spatial and design strategies. The framework applied to Via Filippo Baldinucci in Milan as a case study, where key urban locations such as school, library, park and public transportation stops, were redesigned in terms of the relation with the public spaces by supporting slow mobility. The results demonstrate that prioritizing slow mobility in urban areas leads to significant qualitative improvements. This study offers a holistic analytical and design approach to rethinking urban spaces which demonstrates that slow mobility is an effective tool to make more inclusive, people-centered and livable cities.

1 The Slow E-Mobility Revolution in Urban Planning and Design

The rise and increasing visibility of the e-mobility concept helped the widespread and awareness of the slow mobility concept. The e-mobility revolution is not only transformed modes of transportation but also influenced the understanding of slow mobility. Electric-assisted vehicles facilitated accessibility, even for people who do not have experience on cycling and micromobility. This broadened the perception

A. Gungor (✉) · N. Petaccia
Department of Architecture and Urban Studies, Politecnico Di Milano, Milan, Italy
e-mail: aysegul.gungor@mail.polimi.it

N. Petaccia
e-mail: nicola.petaccia@polimi.it

© The Author(s), under exclusive license to Springer Nature Switzerland AG 2026
R. Moscarelli et al. (eds.), *The Future of Slow Mobility*, Research for Development,
https://doi.org/10.1007/978-3-032-23660-9_10

Fig. 1 Definition of the slow and fast mobility. *Source* Elaborated by the authors

of urban mobility and highlighted its benefits beyond transportation by including health, environmental sustainability, public space design, preventing noise and air pollution. In this context, the question of how slow mobility including slow mood of e-mobility can transform cities is raised. This study focuses on the concept of slow mobility which is also including elements of active mobility (e.g. bikes, scooters, skateboards, people) and e-mobility (e.g. e-bikes and e-scooters), by examining the impact on urban public design and spatial structure (Fig. 1).

Slow mobility includes walking, cycling and other human-powered micromobility modes such as scooters, rollerblades and skateboards that have gained increasing attention in recent years. (Fullagar et al. 2012; Pelgrims 2022; Munch 2024). This category of mobility also extends to e-mobility which includes vehicles, particulary micromobility, powered by electric motors and low-emission technologies (Eldeeb and Tohlob 2022). Collectively, these forms are characterized by low environmental impact and a more human-centered approach to urban movement. Today mobility is not only considered as a technical or transportation issue, but it also stands at the intersection of urgent global challenges which are including climate change, urban inequality and the transformation of everyday life (Hickman and Banister 2007; Chapman 2007; Anderton 2010; Willson 2012) and relevant social changes (Castles 2018). Mobility plays a central role in shaping the future of sustainable liveable cities (Cervero et al. 2017). Slow mobility is directly connected to several United Nations Sustainable Development Goals: Goal 3 (good health and well-being), Goal 7 (affordable and clean energy), Goal 9 (industry, innovation and infrastructure), Goal 11 (sustainable cities and communities), and Goal 13 (climate action) (United Nations 2016; Neun 2018). To align with these goals, cities must be equipped with infrastructure which encourages slow mobility and supports a harmonious integration into the urban life. However, despite this focus, many contemporary planning and design studies on mobility remain broadly technical and often ignore the wider spatial, social and environmental impacts of slow mobility. Especially the lens of the design guidelines, while several of them such as the Swiss manual (Planning Cycle Routes 2008) offer valuable tools for implementing cycle paths and related infrastructure in terms of technical aspects, they often fail to address how mobility shapes public space and everyday life. They should provide strategic visions which help planners in terms of linking mobility infrastructure design with broader urban policy objectives. In that manner, recent studies have begun to explore the relationship between sustainable mobility and urban design and planning by examining how mobility is experienced in everyday urban contexts (Liang et al. 2022). Indeed, studies on urban mobility help to understand the impact of mobility on the social and environmental structures of cities (Gehl 2010; Jacobs 1961).

Beyond existing frameworks, this study aims to reveal that how mobility through its spatial features shapes everyday life and contributes to the dynamics of urban design. The study purposes to address gaps in existing research by focusing on the relationship between slow mobility and public spaces by starting from the spatial needs and dimensional assessment of vehicles and users for both movement and staying activities in the urban environment and concluding spatial and design strategies proposals. The study aims to discuss how slow mobility can transform the design of public spaces and how slow mobility can shape the uses of place. These reflections aim to enrich the current debate by analyzing how planning, design and slow mobility interact and support each other in shaping urban space.

Current urban design practices still often prioritize motorized transport systems by limiting the potential for sustainable mobility and urban quality. Traditional urban design has often disproportionately favoured cars by leading to unequal street usage and contributing to traffic congestion, air pollution, noise and accidents (Gössling et al. 2016). Car users are generally allocated significantly more space than users of other modes, even though they often have shorter travel times (Sasidharan et al. 2023). Studies have shown that car users are allocated approximately 3.5 times more space than non-car users which is including cyclists, scooter riders and pedestrians (Creutzig et al. 2020). In this context, spatial allocation becomes a critical element (Kemmer et al. 2022). Space, particularly in urban mobility, is not only a physical dimension, but it also has public life meaning. From this perspective, mobility is not merely a transportation concern but a spatial phenomenon which is linked to how public space is occupied (Schönfeld and Bertolini 2017). This occupation of space includes both areas dedicated to movement (e.g., cycle paths, sidewalks) and areas for staying (e.g., car parks, squares, bicycle racks). Based on this, a distinction is introduced between slow areas and fast areas which provides an analytical framework to study urban public space. Basically, it investigates how public space is occupied by different forms of mobility which are fast mobility such as cars and slow mobility such as bicycles, walking and stationary needs as parking and engaged activities in public space. Spatial dimensions and the geometric needs of different user groups were measured in order to suggest design strategies and guidelines. The main objective is to explore how the distribution of slow and fast areas influences urban quality, mobility equity and access to collective services. Special attention is given to public spaces that hold collective and symbolic significance such as schools, parks, railway stations, cultural buildings, libraries and places of worship.

Case studies were selected to test and apply these insights. This paper presents the Dergano neighbourhood, located in the northern part of Milan, as an example to apply the framework. Initial results show that mobility habits are influencing how urban spaces are used and designed.

2 Material and Methods

This study investigates mobility mainly as a form of space occupation within urban environments, in addition to being a means of movement. It focuses particularly on slow mobility such as walking, cycling (e.g. traditional bikes, special bikes, e-bikes) and micromobility (e.g. e-scooter, one wheel, hoverboard) and its potential to reshape and reclaim public urban spaces. The methodological framework is structured around two core components, each selected for its contribution to understanding and improving spatial dynamics in the city. Two main components which are following;

- Observing the Spatial Occupation of Slow and Fast Mobility: this step focuses on how different types of mobility occupy urban space. By comparing slow and fast mobility in terms of their spatial requirements during both movement (e.g. sidewalks, bike lanes, roads) and staying (e.g. resting, socializing, parking), the aim is quantifying how much public space is dedicated to each mode. This step is creating a fundamental to identify differences in space allocation and to understand what space can be reclaimed from fast mobility to be reallocated for slow forms.
- Identifying Strategies to Reclaim Public Space: reclaiming space from fast mobility is not enough on its own. The reclaimed spaces should be qualitatively enhanced for urban environments to become more liveable. Therefore, this step focuses on spatial and design interventions which promote usability, comfort, safety and inclusivity. This stage builds on spatial analysis to make them practical, context-sensitive solutions for transforming reclaimed space into quality public environments.

These two steps provide both the analytical foundation which is how space is used and the design direction which is how it can be improved.

2.1 Observing the Spatial Occupation of Slow and Fast Mobility

This section of the research examines spatial occupation through three dimensions: the vehicle or actor (people) itself, the space required during movement and staying. These aspects provide a basis for understanding how different types of mobility shape public spaces. The analysis begins by measuring how much space various modes of mobility occupy and require within the urban environment. Slow mobility is associated with active mobility and e-mobility such as people, bicycles and micromobility vehicles, while fast mobility is represented by the average passenger car. These categories are analyzed in terms of the space they occupy both in movement and staying. Understanding the actual spatial footprint of mobility is essential for evaluating its relationship with public space. Since urban areas are often dominated by fast mobility, recognizing the spatial implications of slow mobility is a key step to create more liveable cities.

2.1.1 The Vehicle or Actor Itself: Space Needed for Vehicles or Actors

The first analytical step concerns the physical dimensions, forms and typologies of the mobility units, namely vehicles and actors. For this study, mobility is categorized into two groups according to their average spatial occupation;

- Slow mobility encompasses three subcategories: people as actors which are including individual pedestrians and those with diverse mobility needs such as a person with children, a person with a stroller and a person with a disability; micromobility vehicles such as monowheels, hoverboards and scooters; bicycles which are including traditional bicycles (e.g. kids bikes, bike sharing, classic bikes, mountain bikes, racing bikes), special types (e.g. tandems, folding bikes, recumbent bikes, handbikes), and cargo bikes (e.g. longtail bikes, cargo tricycle, bucket bikes, commercial cargo); bike extensions such as trailers and tagalong systems are considered.
- Fast mobility is represented in this analysis by the average passenger car which serves as the dominant reference for high occupation mobility in urban contexts.

The comparative framework is based on the two-dimensional ground projection of each unit for slow and fast mobility. The spatial occupation was calculated using the following formula;

$$\mathrm{a} * \mathrm{b} = \mathrm{c}\left(\text{occupation of mobility units m}^2\right)$$

where *a* and *b* reflect the average width and length of the considered mobility unit. The resulting values *c* indicates the mean surface area occupied (Fig. 2).

A comparative table was created to present the average spatial occupancy values across categories within slow mobility (Table 1). The average spatial occupation of a passenger car as a representative of fast mobility is approximately 9 m^2.

2.1.2 During Movement: Space Needed for Circulation

The second analytical step focuses on how much space different types of mobility need during movement with a focus on street design. It compares the space that is used by slow mobility such as pedestrians, bicycles and micromobility vehicles and fast mobility, mainly passenger cars in typical street layouts in this research. A half-street section is used to make a comparison. This means only one side of the street is analyzed. In this layout, a standard vehicle lane for fast mobility is set to 300 cm and is always repeated in the configurations. The rest of the street space is given to slow mobility through different combinations of sidewalks, bike lanes and

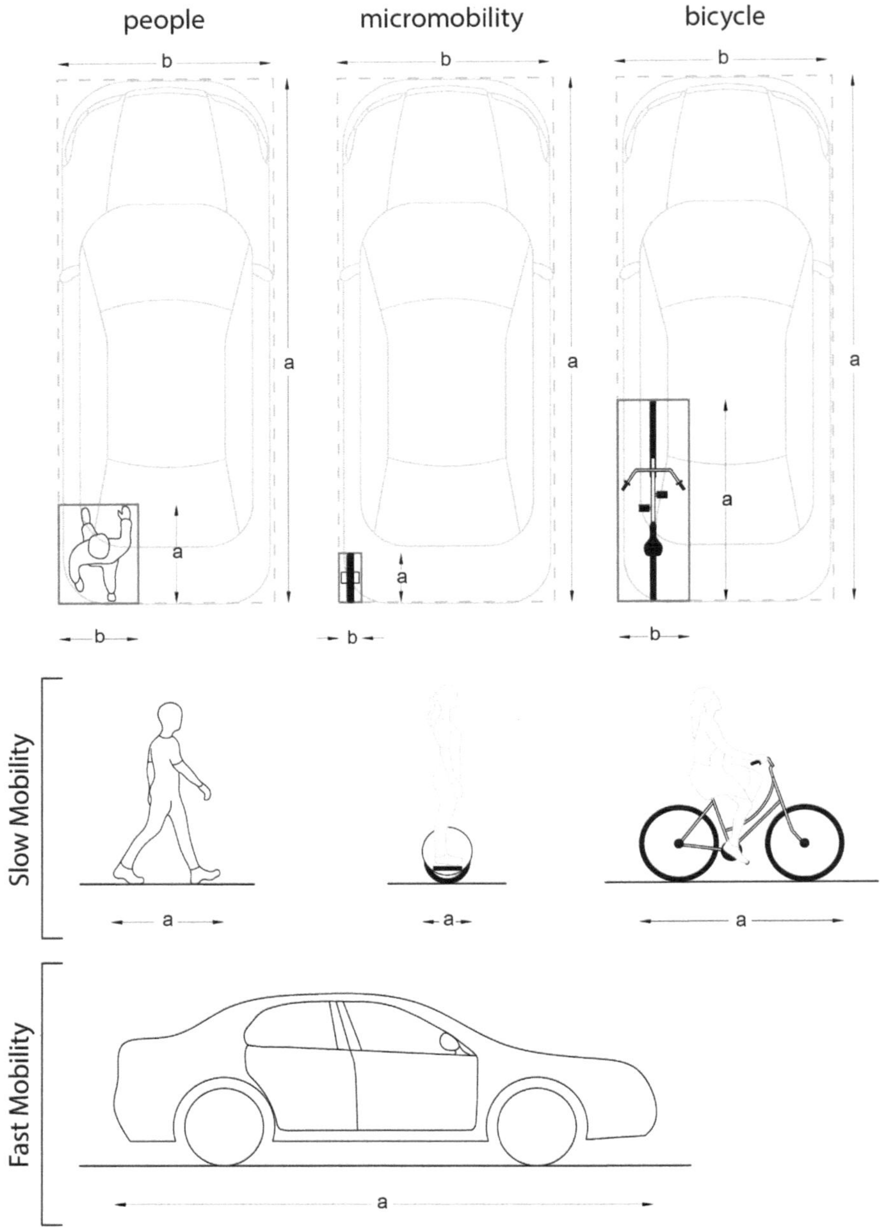

Fig. 2 Comparison of vehicle or actor occupation for slow and fast mobility. Slow mobility includes people, micromobility and bicycle, while fast mobility represents average passenger cars. *Source* Elaborated by the authors

Table 1 Average spatial occupancy of vehicles and actor across categories within slow mobility

Mobility	Category	Type	Occupation (c) [m^2]
Slow mobility	People	One person	0.63
		A person with children	1.08
		A person with a disability	0.86
		A person with a stroller	0.98
	Micromobility and e-micromobility	Monowheel	0.09
		Hoverboard	0.15
		Scooter	0.40
	Bicycle, extensions and e-bike	Kids bikes	0.57
		Racing bikes	0.68
		Classic bikes	1.15
		Bike sharing	1.15
		Mountain bikes	1.15
		Folding bikes	0.75
		Handbike	1.68
		Recumbent bikes	1.68
		Tandem	1.59
		Longtail bikes	1.30
		Cargo tricycle	1.95
		Bucket bikes	1.95
		Commercial cargo	2.70
		Tagalong two wheel	0.68
		Trailer open	0.84
		Trailer close	0.96

micromobility lanes. The street sections are grouped into three main scenarios based on which types of slow mobility they include (Table 2):

- People Only
- Configuration 1: Single vehicle lane; minimum sidewalk
- Configuration 2: Single vehicle lane; average sidewalk
- People, Micromobility and Bicycles
- Configuration 1: Single vehicle lane; shared lane for micromobility and bicycles; average sidewalk
- Configuration 2: Single vehicle lane; separate micromobility lane; separate bicycle lane; average sidewalk
- People and Bicycles Only
- Configuration 1: Single vehicle lane; two-way (dual) bicycle lane; average sidewalk

Table 2 Street section scenarios grouped by types of included slow mobility modes and their configurations

Category	Description	Spatial representation [cm]
People only	Configuration 1: single vehicle lane; minimum sidewalk	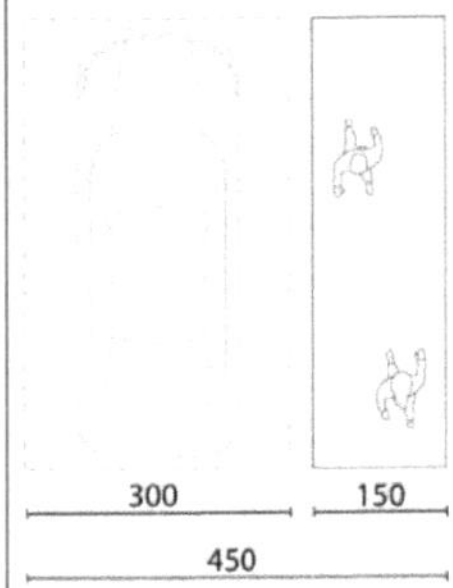
	Configuration 2: single vehicle lane; average sidewalk	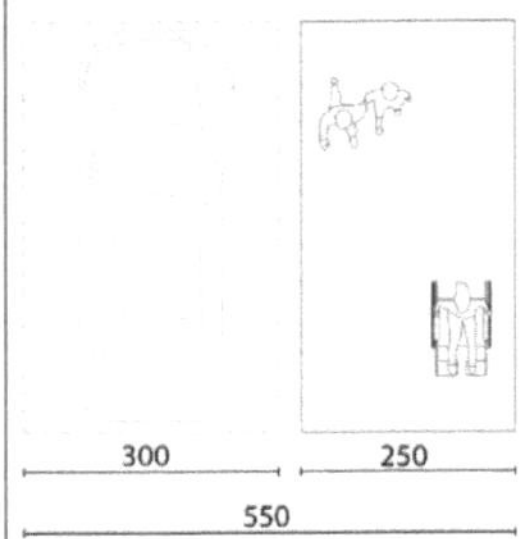
People, micromobility and bicycles	Configuration 1: single vehicle lane; shared lane for micromobility and bicycles; average sidewalk	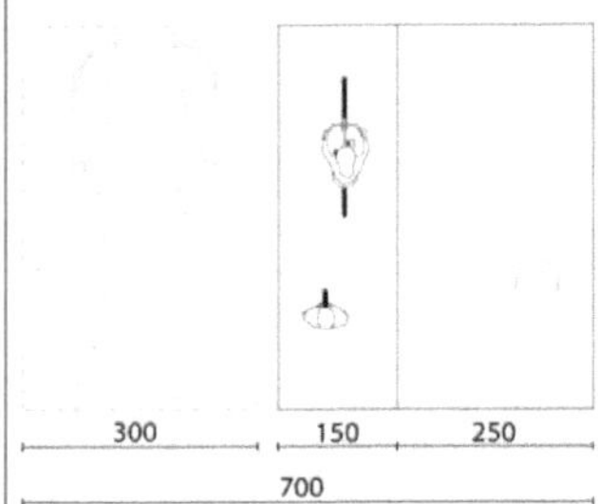

(continued)

Table 2 (continued)

Category	Description	Spatial representation [cm]
	Configuration 2: single vehicle lane; separate micromobility lane; separate bicycle lane; average sidewalk	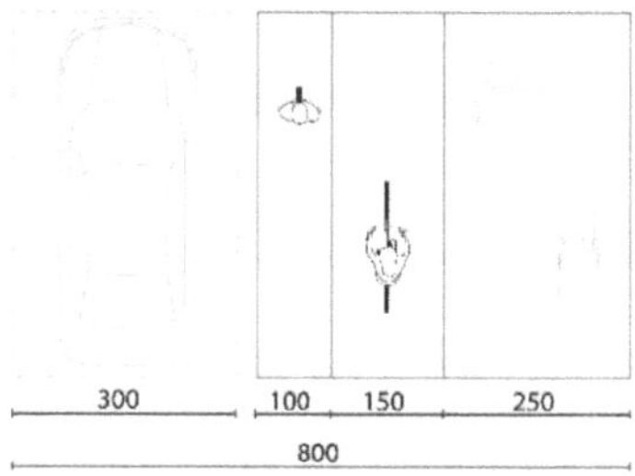
People and bicycles only	Configuration 1: single vehicle lane; two-way (dual) bicycle lane; average sidewalk	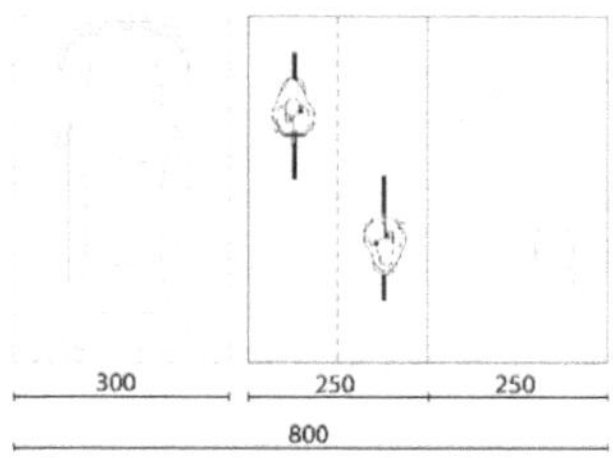

Each of these configurations reveal that how movement space can be used in different ways. Comparing these designs provides a clear picture of how urban streets can be shaped to accommodate different life and usage scenarios.

2.1.3 During Staying: Space Needed Parking and Other Activities

The third analytical step focuses on how much space is required by different mobility modes during the staying phase. This includes both people-centered activities such as resting, socializing, eating or playing and vehicle parking needs. The aim is to understand how public space is used in this phase and to explore opportunities to reclaim the space currently dedicated to fast mobility in favor of more inclusive urban environments through lens of the slow mobility.

The analysis is based on the spatial demands of fast mobility which is represented by the average passenger car parking lot and makes a comparison with those of slow mobility. Fast mobility modes even when stationary occupy significant space. To illustrate, the spatial footprint of a standard passenger car parking space is approximately 12.5 m^2, based on typical dimensions of 250 cm in width and 500 cm in length. This value is used as a baseline reference for further comparisons. In contrast, slow mobility vehicles and people's use of space during the staying phase go beyond only pausing movement. These activities often take place more efficiently in spaces that would otherwise be occupied by fast vehicles, contributing significantly to urban livability. Elements such as benches, pocket parks, sidewalk extensions or small public squares support these functions and when supported by greenery, street furniture, materials and colors greatly enhance the quality of the public realm. This analysis aims to explore the potential uses and benefits of reclaiming a single car parking space for more socially and environmentally valuable purposes. The analysis addresses two main aspects to show alternative uses of a standard car parking space (Table 3);

- People and the Use of Space: design examples demonstrate how space traditionally allocated for car parking can be repurposed such as green and social areas with trees and benches; small-scale play areas for children; outdoor seating for cafes or food services.
- Parking Requirements; various bicycle and micromobility parking configurations were analyzed to highlight spatial efficiency such as linear perpendicular bike racks up to 10 bicycles, linear diagonal bike racks up to 8 bicycles, two-tier bike racks up to 12 bicycles, covered or box bike racks up to 5 bicycles, micromobility parking areas which are suitable for vehicles such as e-scooters, monowheels or hoverboards.

These examples emphasize how a single car parking space can serve multiple users when reallocated for slow mobility. This supports the broader argument for reclaiming car-dominated urban areas in favor of more flexible, people-centered public spaces.

2.2 *Identifying Strategies to Reclaim Public Space*

Following the analysis of spatial occupation by slow and fast mobility, this second phase examines strategies to reclaim public space from fast mobility and improve its quantitative and qualitative characteristics for liveability. The analysis is articulated through three main sections:

- spatial strategies: gaining space from fast mobility;
- design strategies: creating higher quality environments;
- living the spaces of slow mobility in key urban locations

A distinction is made between spatial and design strategies. This separation reflects two related but distinct dimensions of reclaiming public space. Spatial strategies focus on the geometric transformation of urban areas by reducing or reconfiguring spaces previously dominated by fast mobility by providing more space for slow mobility. This step is concerned with reclaiming physical surface area for walking, cycling and micromobility both at the moment of movement and staying. On the other hand, design strategies reflect the qualitative enhancement of these reclaimed spaces. Gaining space is not sufficient alone. They must be equipped with appropriate materials, furniture, greenery, vegetation and elevation treatments. In this sense, the methodological framework combines quantitative spatial reallocation to qualitative

Table 3 Alternative uses of a standard car parking space

Category	Description	Spatial representation [cm]
People	Green and social areas	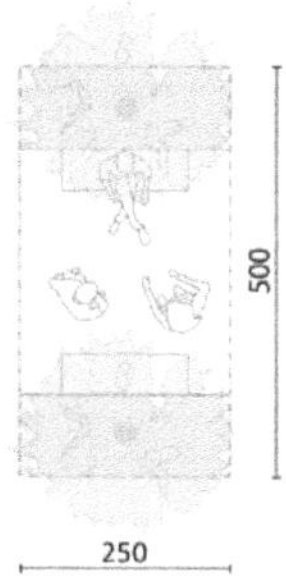
	Small-scale play areas for children	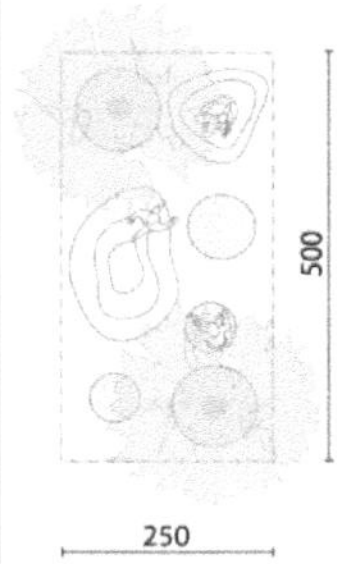
	Outdoor seating for cafes or food services	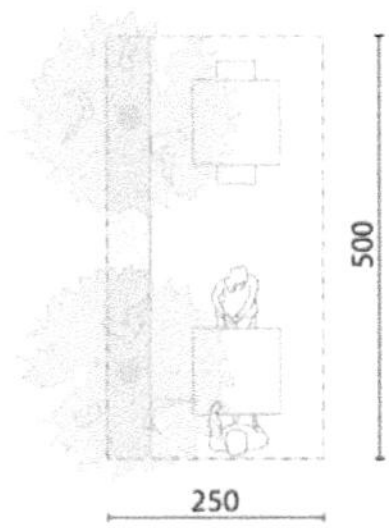
Parking requirements	Linear perpendicular bike racks up to 10 bicycles	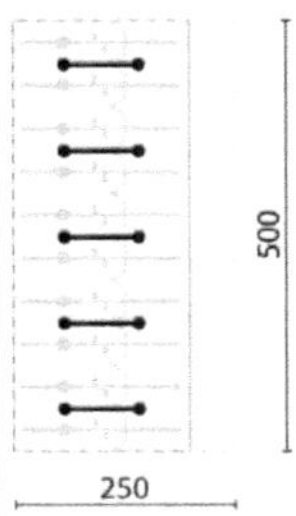

(continued)

Table 3 (continued)

Category	Description	Spatial representation [cm]
	Linear diagonal bike racks up to 8 bicycles	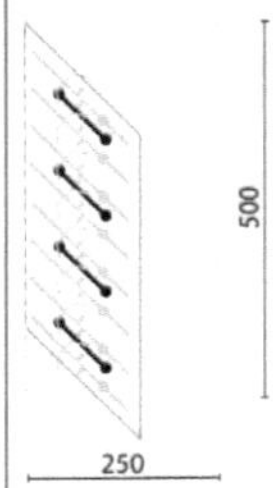
	Two-tier bike racks up to 12 bicycles	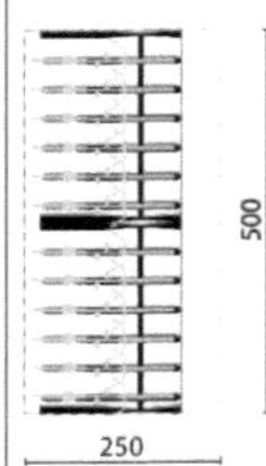
	Covered or box bike racks up to 5 bicycles	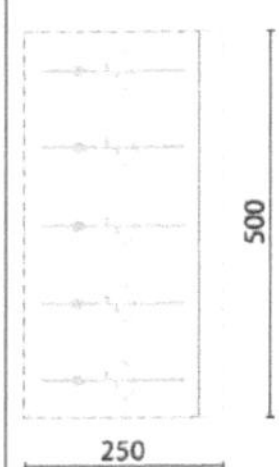
	Micromobility parking areas	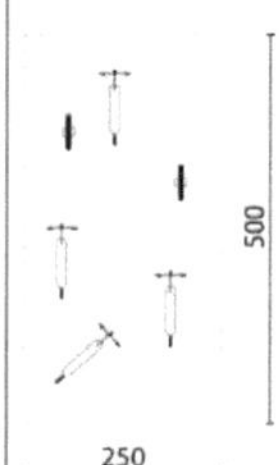

spatial improvement. First, occupied space is physically reclaimed, then it is designed by supporting everyday activities, promotes safety and responds to the diverse needs of urban users. The combined application of these two strategies to selected key urban locations not only initiates but also completes the process of public space reclaiming. Building on this objective, the third stage of the analysis concretizes the strategies in the context of specific urban functions and user needs.

2.2.1 Spatial Strategies: Gaining Space from Fast Mobility

The initial phase of interventions focuses on spatial strategies that quantitatively transform urban space by physically reducing the dominance of fast mobility and reallocating the reclaimed surface area to slow mobility. These spatial strategies involve two approaches:

- Reduction: this includes narrowing vehicle lanes by eliminating on-street parking or some lanes and reallocating the space to pedestrians, cyclists and micromobility users.
- Reconfiguration: modifying street geometry through techniques such as curb extensions, pinch points and shared streets (e.g., woonerfs) which reduce vehicular priority and promote slower and more inclusive mobility.

2.2.2 Design Strategies: Creating Higher Quality Environments

Following the spatial strategies, design strategies focus on the qualitative improvement of these spaces to enhance usability, comfort, safety and inclusivity. Key components include:

- Urban Furniture: the integration of benches, play equipment, bike racks and other elements which support resting, socializing, playing, eating and parking needs.
- Greenery: integrating green such as trees, planting beds and depaved areas to enhance microclimate, biodiversity and aesthetic quality.
- Material and Colors: applying differentiated surface materials and colors to identify functions such as walking and cycling paths by improving safety and orientation.
- Elevation Treatments: utilizing raised pedestrian crossings and level surfaces to slow down fast vehicles and facilitate barrier-free and accessible movement for all users.

These design interventions contribute to transforming reclaimed spaces into vibrant, inviting and multifunctional public areas.

2.2.3 Living the Spaces of Slow Mobility in Key Urban Locations

This final phase, based on spatial occupation analysis and spatial and design strategies, focuses on how reclaimed urban spaces are experienced and shaped by slow mobility in everyday urban life. While previous sections quantified the space occupied by different forms of mobility and proposed ways to reclaim and improve these spaces, this part shifts its focus to real world applications of these principles. It examines how spatial and design interventions interact with specific urban functions by responding to diverse rhythms and users. The process of reclaiming space from lens of slow mobility is not a single solution. Public space transformation must be context sensitive by reflecting the diverse spatial, social and infrastructural functions of the urban fabric (Gehl 2011). To illustrate this, four key dimensions of urban space are introduced to understand the key urban locations. Each of them addresses different needs and user profiles within the framework of slow mobility:

- Daily Activity Dimension: these are spaces where daily routines occur such as schools, parks, supermarkets and sports centers. These environments are typically located in residential or mixed-use areas where people can arrive on foot, by bike or using micromobility. Here, spatial strategies are considered, especially in conjunction with the daily rhythm and taking into account the user profile such as children, families and the elderly.
- Social Dimension: this category includes museums, cultural centers, theaters and worship places that host both formal events and informal gatherings specific time and days. Spatial interventions are considered to meet the needs of gathering and density at certain hours.
- Civil Dimension: buildings such as libraries, municipal offices and public service centers serve a wide range of users including students, employees, and the elderly.
- Infrastructure Dimension: transportation nodes such as bus stops, train stations and metro stations represent interfaces between slow and fast mobility.

3 Results

The methodology was applied to a selected site in Milan, 250 m section of Via Filippo Baldinucci. Street segment located in the Dergano district of northern Milan. This area is mixed-use residential area and far from Milan's main core. Six key urban locations which include a school, library, public offices, park, transportation stop and supermarket are located along the street. These key urban locations shape the social and cultural life of the street. Firstly, the existing situation of the street segment is analyzed. Initial research revealed that several issues affecting accessibility, safety and the daily and community-oriented functions of the area. These included insufficient space and visibility in front of the school, car parking cutting across the slow mobility area, bus stops and bike racks obstructing pedestrian flow, intersections do not prioritize pedestrians, squares dominated by cars in front of the library and supermarket and public areas lacking comfort, greenery and legibility. The spatial distribution of mobility was nearly balanced, with slow mobility occupying 51% of the street and fast mobility 49% (Fig. 3).

To address these problems, a series of targeted spatial and design strategies were implemented which are reduction of car parking, repositioning of bus stops and bike racks, reconfiguration of intersections, creation of new civic and commercial squares and enhancements to public areas including urban furniture, greenery, raised crossings and differentiated materials and colors. As a result of these interventions, the spatial distribution of mobility shifted significantly, with slow mobility increasing to 80% and fast mobility decreasing to 20%. (Fig. 4).

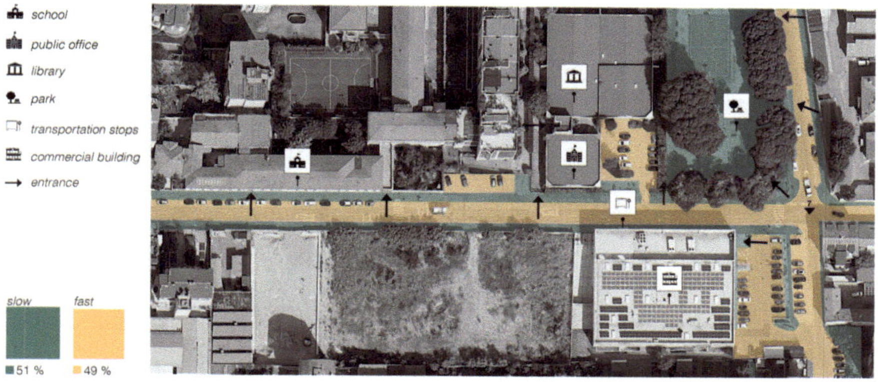

Fig. 3 Allocation of current slow and fast mobility. *Source* Elaborated by the authors

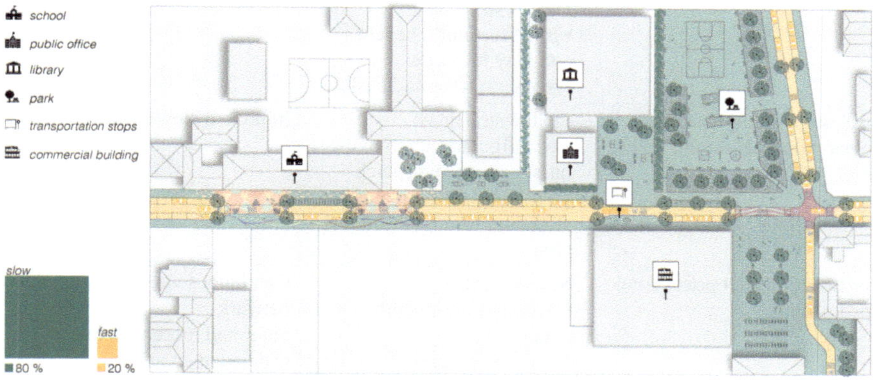

Fig. 4 Allocation of mobility after targeted interventions by reflecting spatial and design strategies. *Source* Elaborated by the authors)

Qualitative benefits include safer and more visible school access, improved pedestrian and micromobility flow, enhanced civic identity of the library and public office, revitalized supermarket square and the creation of a continuous network linking all six key urban locations. These outcomes collectively enhance the liveability, inclusiveness and people-oriented potential of public space in this section of Via Filippo Baldinucci (Table 4).

Overall, the interventions demonstrate that targeted spatial and design strategies tailored to the needs of key locations have yielded both quantitative and qualitative improvements. Slow mobility increased from 51 to 80%, while safer school access, improved pedestrian flow and revitalized public spaces have strengthened the street's liveability, inclusiveness and people-oriented character.

Table 4 Identified problems, applied spatial and design strategies, and resulting outcomes

Problem identified	Strategy applied	Outcome
Insufficient space and visibility in front of the school	Spatial strategies: • reconfiguration; removing on street parking (pinch point) Design strategies: • urban furniture: bike racks •greenery: trees • material and colors: colored in front of the school pavement • elevation treatment: raised pedestrian crossing	Safer and more visible school access; improved pedestrian experience
Car parking cutting across slow mobility area	Spatial strategies: •reduction: removing car parking lots Design strategies: •urban furniture: benches and play furniture •greenery: trees and bushes	Slow mobility space expanded and made uninterrupted
Bus stops and e-bike racks obstructing pedestrian flow	Spatial strategies: •reconfiguration: removing on-street parking (pinch point). Repositioning of bus stops and bike racks on this area	Improved slow mobility flow
Intersections not prioritizing pedestrians	Spatial strategies: •reconfiguration: curb extensions Design Strategies: •material and colors: colored pedestrian crossing •elevation treatments: raised pedestrian crossing	Continuous slow mobility network Safer and more visible pedestrian crossing
Squares in front of library and supermarket dominated by cars	Spatial strategies: •reduction: removing car parking lots Design Strategies: •urban furniture: benches, play furniture, bike racks • greenery: trees, bushes and grass • material and colors: different pedestrian paving for square	Revitalized public squares; enhanced civic identity of library and public office; more inviting commercial spaces

4 Conclusion

This study demonstrates that analyzing mobility through the lens of spatial occupation reveals great opportunities to reclaim and transform public spaces through slow mobility. By adopting a two-step methodological framework, the first quantifies the spatial needs of slow and fast mobility, while the second applies spatial and design strategies to reallocate and develop urban space. The research offers both analytical and practical recommendations. Spatial occupation analysis has shown that fast mobility occupies a large portion of public space, particularly during "staying" phases such as parking. In contrast, slow mobility which encompasses walking, cycling and micromobility, particulary slow e-mobillity, offers significant spatial efficiency by taking up less space. This is an important element to reclaim spaces for public life. The empirical application of this framework to Milan, Via Filippo Baldinucci demonstrates the transformative potential of slow mobility. The integration of design strategies such as greenery, street furniture, differentiated materials and colors and elevation treatments into the spatial strategies has led to measurable changes. For example, in the case study, the increase in slow mobility from 51 to 80% not only supply more spaces but also increased the quality of the urban spaces. Through this dual approach which is combining data and design, the study contributes to current discussions on car-dependent cities by highlighting how slow mobility practices shape urban spaces. Finally, while many existing studies focus on producing technical guidelines for mobility solutions, this research aim goes beyond an only technical exercise by integrating spatial and design strategies with the key urban locations which respond to everyday urban life.

Author contributions and acknowledgements

The authors have shared the concept of the paper. A.G. has written the paragraphs 2, 3. N.P. has written the paragraphs 1. Together have written the paragraphs 4.

The part written by N.P. is a result of the study carried out within the MOST—Sustainable Mobility National Research Center and received funding from the European Union Next-GenerationEU (PIANO NAZIONALE DI RIPRESA E RESILIENZA (PNRR)—MISSIONE 4 COMPONENTE 2, INVESTIMENTO 1.4—D.D. 1033 17/06/2022, CN00000023). This manuscript reflects only the author' views and opinions, neither the European Union nor the European Commission can be considered responsible for them.

References

Anderton K (2010) Two billion cars: driving towards sustainability. Transp Rev 30(6):675–676

Castles S (2018) Social transformation and human mobility: reflections on the past, present and future of migration. J Intercult Stud 39:238–251

Cervero R, Al S, Guerra E (2017) Beyond mobility: planning cities for people and places. Island Press, Washington, DC. https://doi.org/10.5822/978-1-61091-835-0

Chapman L (2007) Transport and climate change: a review. J Transp Geogr 15(5):354–367. https://doi.org/10.1016/j.jtrangeo.2006.11.008
Creutzig F, Javaid A, Soomauroo Z, Lohrey S, Milojevic-Dupont N, Ramakrishnan A (2020) Fair street space allocation: ethical principles and empirical insights. Transp rev 40(6): 711–733
Eldeeb A, Tohlob A (2022) Waterfront slow-motility development (Mansoura—Egypt). Eng Res J 45 (3):493–506. https://doi.org/10.21608/erjm.2022.135831.1171
Fullagar S, Markwell K, Wilson E, (2012) Starting slow: thinking through slow mobilities and experiences. In: Fullagar S, Markwell K, Wilson E (eds) Slow tourism. Experience and Mobility, Multilingual Matters & Channel View, Bristol, p 1–9
Gehl J (2010) Cities for people. Island Press, Washington D.C
Gehl J (2011) Life between buildings. Island Press, Washington D.C, Using Public Space
Gössling S, Schröder M, Späth P, Freytag T (2016) Urban space distribution and sustainable transport. Transp Rev 36(5):659–679
Hickman R, Banister D (2007) Looking over the horizon: transport and reduced CO_2 emissions in the UK by 2030. Transp Policy 14(5):377–387
Jacobs J (1961) The death and life of great American cities. Random House, New York
Kemmer L, Sgibnev W, Weicker T, Woods M (2022) Spaces of exposure: re-thinking 'publicness' through public transport. Cult Geogr 29(2):285–299. https://doi.org/10.1177/14744740211068097
Liang Y, D'Uva D, Scandiffio A, Rolando A (2022) The more walkable, the more livable?— can urban attractiveness improve urban vitality? Transp Res Procedia 60:322–329. https://doi.org/10.1016/j.trpro.2021.12.042
Munch E (2024) Sustainable and slow cities: the quest for conviviality. Sustain: Sci Pract Policy 20(1):1–14. https://doi.org/10.1080/15487733.2024.2375810
Neun M (2018) Framing sustainable mobility in practice and research—and rethinking 'reverse innovation' in the case of active mobility. In: Grafl K, Bunte H, Dziekan K, Haubold H, Neun M (eds) framing the third cycling century—bridging the gap between research and practice. UBA German Environment Agency: Dessau-Roßlau, Germany, ECF European Cyclist Federation, Brussels, Belgium
Pelgrims C (2022) Towards ethical mobility behavior: the socio-ecological potentiality of ambiance-based mobility infrastructure design. In: Pelgrims C (ed) Structures and architecture a viable urban perspective? CRC Press, Boca Raton, pp 527–532
Planning Cycle Routes—Manual (2008) Ed. Federal Roads Office (FEDRO) and Swiss Cycling. https://www.newsd.admin.ch/newsd/message/attachments/11128.pdf. Accessed 10 June 2025
Sasidharan M, Torbaghan ME, Fathy Y, Rogers CD, Metje N, Schooling J (2023) Designing user-centric transport strategies for urban road space redistribution. Commun Transp Res 3:100109. https://doi.org/10.1016/j.commtr.2023.100109
United Nations (2016) Mobilizing sustainable transport for development: analysis and policy recommendations from the United Nations secretary-general's high-level advisory group on sustainable transport. https://sustainabledevelopment.un.org/index.php?page=view&type=400&nr=2375&menu=1515. Accessed 15 Sep 2025
Von Schönfeld KC, Bertolini L (2017) Urban streets: epitomes of planning challenges and opportunities at the interface of public space and mobility. Cities 68:48–55. https://doi.org/10.1016/j.cities.2017.04.012
Willson R (2012) Transport revolutions: moving people and freight without oil. J Urban Technol 19:106–107

Sustainable Logistics: The Impact of e-Bikes in Last Mile Delivery

Paolo Beatrici, Francesca Maggioni, Sebastian Birolini, and Paolo Malighetti

Abstract In recent years, urban freight distribution has been deeply reshaped by rapid urbanization, the exponential growth of e-commerce, and evolving consumer expectations. The Last Mile (LM) segment, widely recognized as the most costly and inefficient part of the supply chain, presents critical challenges in terms of cost-effectiveness, operational efficiency, and environmental sustainability. One promising solution involves the integration of light and electric vehicles, particularly electric cargo bikes, into urban delivery fleets. This chapter provides a review of the literature on sustainable and light urban mobility, with a specific emphasis on modeling the use of electric cargo bikes in the context of Vehicle Routing Problems (VRPs) and Fleet Size and Mix VRPs (FSMVRPs) applied to Last Mile logistics. Given the inherent uncertainties in these operations, such as variable customer demand, travel and service times, and energy consumption, special attention is given to stochastic optimization approaches. The chapter concludes by highlighting key gaps in the literature and proposing directions for future research.

1 Introduction

The evolution of urban logistics has been profoundly influenced by the rapid pace of urbanization, the widespread adoption of e-commerce, and the growing societal emphasis on sustainability. The increasing volume of online orders has led to a

P. Beatrici · F. Maggioni (✉) · S. Birolini · P. Malighetti
Department of Management, Information and Production Engineering, University of Bergamo, Bergamo, Italy
e-mail: francesca.maggioni@unibg.it

P. Beatrici
e-mail: paolo.beatrici@unibg.it

S. Birolini
e-mail: sebastian.birolini@unibg.it

P. Malighetti
e-mail: paolo.malighetti@unibg.it

© The Author(s), under exclusive license to Springer Nature Switzerland AG 2026

R. Moscarelli et al. (eds.), *The Future of Slow Mobility*, Research for Development,
https://doi.org/10.1007/978-3-032-23660-9_11

significant expansion in urban transport operations (Shahmohammadi et al. 2020), which, combined with the dominant reliance on Conventional Vehicles (CVs), has contributed to a range of urban challenges, including traffic congestion, greenhouse gas emissions, and both air and noise pollution. As the Last Mile (LM) is widely acknowledged as one of the most complex and least efficient segments of the supply chain (Mohammad et al. 2023), logistics providers are under mounting pressure to rethink their delivery strategies in order to mitigate environmental impacts and contribute to more livable urban environments.

One particularly promising approach involves the integration of Light Freight Vehicles (LFVs) into urban delivery fleets. LFVs offer the potential to reduce emissions and ease traffic congestion while maintaining high standards of delivery efficiency. Among these, Cargo Bikes (CBs) and Electric Cargo Bikes (ECBs) have emerged as especially viable alternatives. According to Wrighton and Reiter (2016), up to 51% of commercial and private motorized trips in urban areas could feasibly be replaced by cargo bikes, underscoring their capacity to balance operational performance with environmental benefits (Blazejewski et al. 2020). Additionally, a study conducted in Nocerino et al. (2016) demonstrates that the deployment of an electric fleet, despite slightly higher salary costs, achieves nearly identical overall costs while drastically reducing both energy expenditures and CO_2 emissions, thereby confirming ECBs as a viable and sustainable alternative for Last Mile delivery operations in urban areas.

The field of Operations Research has actively contributed to the advancement of sustainable LM delivery, promoting the incorporation of environmental objectives across various stages of the distribution process. In this context, numerous optimization and simulation models have been developed to address the complexities of light and sustainable urban mobility, frequently through the formulation of Vehicle Routing Problems (VRPs) and their many variants.

The purpose of this article is to review the contributions from the Operations Research literature related to light and sustainable mobility, with a specific focus on the role of light vehicles, particularly electric cargo bikes, in VRP variants. The reviewed literature is categorized along the following dimensions:

- **Fleet structure**: distinguishing between studies that assume a fixed delivery fleet and those that also consider tactical decisions regarding fleet size and composition;
- **Integration of ECBs**: differentiating works based on whether or not electric cargo bikes are included in the delivery fleet.

Furthermore, given the multiple sources of uncertainty that characterize urban LM logistics, such as demand variability, travel times, and service durations, this review also examines stochastic optimization approaches that have been proposed in this context.

The chapter is structured as follows: Sect. 2 outlines the key modeling features of VRP and its main variants as applied to sustainable urban logistics. Section 3 presents a state-of-the-art literature review on sustainable LM delivery, with particular attention to the role of ECBs. Finally, Sect. 4 summarizes the main findings and outlines potential directions for future research.

2 Modeling Features of Sustainable Last-Mile Delivery Problems

This section presents the main modeling features associated with sustainable LM delivery problems. In the field of Operations Research, these problems are typically formulated as VRPs or one of their many variants. We identify two principal modeling approaches used to assess the impact of LFVs and ECBs on LM logistics. Figures 1 and 2 illustrate examples of delivery networks based on these two approaches.

In the first approach, a fixed fleet of homogeneous or heterogeneous vehicles is stationed at a central distribution center. Routing decisions are made during the operational phase to optimize delivery routes for the available vehicles (see Fig. 1).

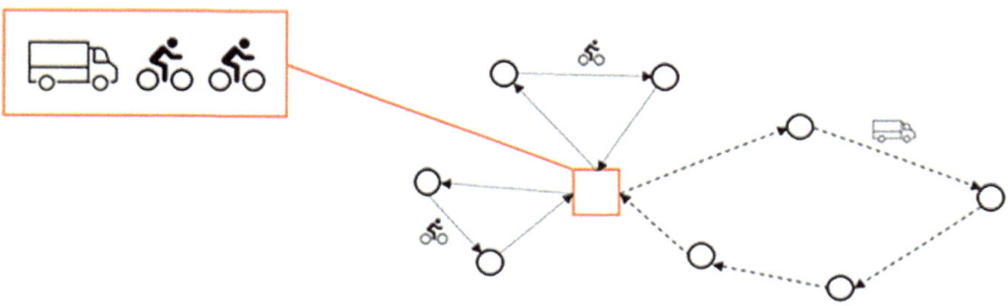

Fig. 1 Example of a LM delivery network with one central depot (orange square) and eight customers (black nodes). A fixed fleet comprising three heterogeneous vehicles (one CV and two ECBs) is based at the depot. Routing plans are developed to assign delivery routes to each vehicle

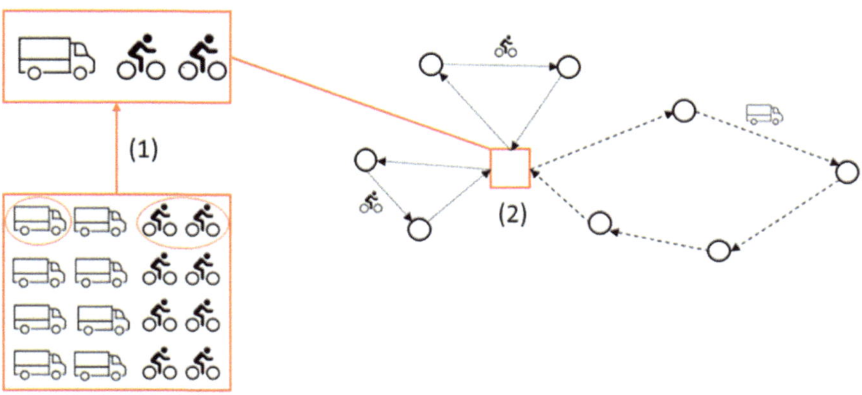

Fig. 2 Example of a LM delivery network with one central depot (orange square) and eight customers (black nodes). In the tactical phase (1), the fleet configuration (e.g., one CV and two ECBs) is determined based on factors such as customer demand or delivery volume. In the subsequent operational phase (2), routing decisions are made for the selected fleet

This modeling framework forms the basis of classical VRPs and several well-known variants, such as the Heterogeneous VRP (HVRP), the Green VRP (GVRP), and the Two-echelon VRP (2eVRP).

The second approach involves making tactical decisions regarding the size and composition of the fleet prior to the operational phase (see Fig. 2). This framework is central to the Fleet Size and Mix VRP (FSMVRP).

We now introduce the common notation to VRPs and their variants (see Koç et al. 2016): let $\mathcal{G} = \{\mathcal{I}, \mathcal{A}\}$ be a complete graph, where $\mathcal{I} = \{0, \ldots, N\}$ is the set of nodes and $\mathcal{A} = \{(i, j) : 0 \leq i, j \leq N, i \neq j\}$ is the set of arcs. Node 0 represents the starting depot and N the ending depot (possibly a duplicate of 0). The distance between nodes i and j is denoted by d_{ij}. The customers set is $\mathcal{I}_{0,N} = \mathcal{I}\setminus\{0, N\}$. Each customer $i \in \mathcal{I}_{0,N}$ has a non-negative demand d_i. Sets $\mathcal{I}_0$ and $\mathcal{I}_N$ denote $\mathcal{I}\setminus\{0\}$ and $\mathcal{I}\setminus\{N\}$, respectively. Let $\mathcal{P} = \{1, \ldots, P\}$ denote the set of available vehicle types. Each vehicle type $p \in \mathcal{P}$ has a load capacity l_p, a cost c_p, and traversal cost c_{ijp}. Binary decision variables x_{ijp} indicate whether arc (i, j) is used by a vehicle of type p. The objective is to satisfy customer demand while minimizing total delivery costs, ensuring that vehicle capacities are not exceeded.

We now outline the distinguishing characteristics of each VRP variant considered in this review.

Heterogeneous Vehicle Routing Problem (HVRP) HVRPs involve a fixed fleet with a predetermined size and composition. The objective function is given by:

$$\sum_{p \in \mathcal{P}} \sum_{j \in \mathcal{I}_0} c_p x_{0jp} + \sum_{p \in \mathcal{P}} \sum_{i \in \mathcal{I}_N} \sum_{j \in \mathcal{I}_0} c_{ijp} x_{ijp}, \tag{2.1}$$

where the first term represents the fixed cost of deploying vehicle of type p, and the second term captures the routing cost. A fleet size constraint is imposed using parameter m_p as follows:

$$\sum_{j \in \mathcal{I}_0} x_{0jp} \leq m_p, \quad p \in \mathcal{P}. \tag{2.2}$$

Green Vehicle Routing Problem (GVRP) GVRPs extend VRPs by integrating environmental metrics (e.g., greenhouse gas emissions, noise, and air pollution) into the objective function. While the basic structure mirrors (2.1), these models often include multi-objective criteria or uncertainty modeling. For a detailed taxonomy, we refer to Moghdani et al. (2021).

Fleet Size and Mix VRP (FSMVRP) In the FSMVRP, both fleet sizing and routing decisions are made concurrently. Here, c_p reflects the acquisition cost of vehicles rather than a fixed ownership cost. Constraints on the number of vehicles (2.2) are omitted, as the fleet is to be determined within the model.

Two-Echelon VRP (2eVRP) The 2eVRP involves a two-tier distribution system: goods are transported from a central depot to satellite depots, from which smaller vehicles (e.g., ECBs) complete the final delivery leg. This hierarchical structure supports sustainable vehicle usage in the second echelon. For comprehensive formulations, see Perboli et al. (2011).

2.1 Additional Features in Sustainable LM Delivery

Beyond standard VRP constraints (e.g., customer visits, capacity limits, subtour elimination, see Toth and Vigo 2014), specific constraints related to LFVs and ECBs must be considered:

Time Windows Time-sensitive deliveries are essential in LM logistics, especially when using LFVs and ECBs, which often operate at lower speeds than conventional vehicles. To account for delivery timing requirements, each customer node i is associated with a service duration s_i and a time window $[a_i, b_i]$, within which service must start.

Let τ_i denote the service start time at node i and t_{ijp} the travel time between nodes i and j for a vehicle of type p. The time window constraints can be expressed as:

$$a_i \le \tau_i \le b_i, \quad i \in \mathcal{I}_{0,N}. \tag{2.3}$$

To ensure temporal consistency between consecutive visits, the following constraint enforces that the service at node j starts only after completing travel and service at node i:

$$\tau_i + \left(t_{ijp} + s_i\right)x_{ijp} - b_0\left(1 - x_{ijp}\right) \le \tau_j, \quad i \in \mathcal{I}_0, j \in \mathcal{I}_N \backslash \{i\}, p \in \mathcal{P}. \tag{2.4}$$

Recharging Operations ECBs, like other electric vehicles, have limited battery capacity, which constrains their operational range. To ensure continuity of service, delivery routes must incorporate the possibility of recharging stops.

Let $\mathcal{F}$ be the set of charging stations, and $\mathcal{F}'$ a set of replicated nodes allowing multiple visits to the same station.

Each vehicle type $p \in \mathcal{P}$ is defined by its battery capacity e_p, energy consumption rate r_p, and recharging time g_p. Let y_{ip} represent the battery level of a vehicle of type p at node i. Battery constraints ensure that vehicles do not exceed their energy capacity.

Two formulations are typically used, depending on the nature of the departure node:

- From regular customer or transit nodes:

$$0 \le y_{jp} \le y_{ip} - \left(r_p d_{ij}\right)x_{ijp} + e_p\left(1 - x_{ijp}\right), \quad i \in \mathcal{I}_{0,N}, j \in \mathcal{I}_0 \cup \mathcal{F}_{0,N'} \backslash \{i\}, p \in \mathcal{P}. \tag{2.5}$$

The term $e_p(1 - x_{ijp})$ allows the constraint to remain valid even when arc (i, j) is not used.

- From recharging stations:

$$0 \leq y_{jp} \leq y_{ip} - (r_p d_{ij}) x_{ijp}, \quad i \in \mathcal{F}'_N, j \in \mathcal{I}_0 \cup \mathcal{F}_{0,N'} \backslash \{i\}, p \in \mathcal{P}. \tag{2.6}$$

In this case, the arc is assumed to be active, so the relaxation term is omitted. Initial battery level at the depot is set to full capacity:

$$y_{0p} = e_p, \quad p \in \mathcal{P}. \tag{2.7}$$

To incorporate recharging time into the time constraints, the following condition adjusts the schedule accordingly:

$$\tau_i + t_{ijp} x_{ijp} + g_p(e_p - y_{ip}) - (b_0 + g_p e_p)(1 - x_{ijp}) \leqslant \tau_j, \\ i \in \mathcal{F}_{0,N'}, j \in \mathcal{I}_0 \cup \mathcal{F}_{0,N'} \setminus \{i\}, p \in \mathcal{P}. \tag{2.8}$$

Finally, constraint (2.4) is modified when arc (i, j) originates from a recharging station:

$$\tau_i + (t_{ijp} + s_i) x_{ijp} - b_0(1 - x_{ijp}) \leq \tau_j, \quad i \in \mathcal{I}_N, j \in \mathcal{I}_0 \cup \mathcal{F}_{0,N'} \backslash \{i\}, p \in \mathcal{P}. \tag{2.9}$$

Sources of Uncertainty in LM Delivery Another critical aspect affecting the performance of LM delivery systems is the presence of various sources of uncertainty across key operational parameters:

- *Customer demand*: the actual quantity of goods to be delivered may deviate from forecasts, making it difficult to allocate the appropriate fleet size. This can lead to unfulfilled deliveries and necessitate costly returns to the depot for restocking.
- *Customer location*: in some circumstances, delivery addresses may only be disclosed at the start of or during operations. This lack of prior knowledge can challenge route planning, especially when vehicle capacity or range is limited. In such cases, fallback strategies like split deliveries or outsourcing to third-party providers may be necessary.
- *Energy consumption*: the energy usage of electric vehicles, including ECBs, is affected by multiple external factors such as weather conditions, terrain, and payload. These fluctuations may reduce the effective range of the vehicle and necessitate additional charging stops mid-route (Bezzi et al. 2023, Spinelli et al. 2026)

To manage these challenges, stochastic and robust optimization techniques provide a framework capable of delivering solutions that perform well across a range of uncertainty realizations or scenarios (see e.g. Birge and Louveaux 2011; King and Wallace 2012). Nevertheless, as will be discussed in Sect. 3, its application in the context of sustainable LM logistics is still relatively underexplored, highlighting promising directions for future research.

3 Sustainable Last Mile Delivery Problems

This section reviews contributions from the Operations Research literature concerning the integration of LFVs and ECBs in urban logistics systems. The reviewed studies are organized into three categories:

- **Section 3.1** covers VRP variants assuming a fixed fleet, where only operational routing decisions are taken.
- **Section 3.2** focuses on VRP variants that also consider tactical decisions regarding fleet size and composition, typically modeled as FSMVRPs.
- **Section 3.3** specifically examines the role of ECBs in sustainable LM delivery.

At the end of each subsection, we briefly summarize the solution methods employed. While a comprehensive methodological review is beyond the scope of this article, we indicate whether exact or heuristic approaches are used and mention the main techniques adopted.

Table 1 provides an overview of all referenced works. It includes the modeling approach (optimization or simulation), fleet characteristics (e.g., fixed, heterogeneous, use of ECBs), problem type, presence of uncertainty, and the solution method adopted (exact, heuristic and type of algorithm).

3.1 Sustainable LM Delivery Problems with Fixed Fleet

A significant number of studies have focused on operational routing decisions under the assumption of a fixed vehicle fleet, typically within a deterministic framework, as summarized in Table 1.

The role of LFVs, such as small electric vehicles, is addressed in Schneider et al. (2014), where a homogeneous fleet is used to provide green delivery services with time window constraints. Two-level distribution systems combining LFVs and CVs are examined in Anderluh et al. (2017), which introduces a 2eVRP with temporal and spatial synchronization between vans and cargo bikes. A similar modeling approach is found in Mühlbauer and Fontaine (2021), who consider asymmetric distances for cargo cycles. Additional 2eVRP extensions include the use of parcel lockers in Enthoven et al. (2020), and the integration of electric vehicles with battery swapping stations in Jie et al. (2019).

Mixed-fleet routing is also explored in Ćirović et al. (2014), where a GVRP is proposed to support sustainable urban deliveries using both CVs and LFVs. Goeke and Schneider (2015) address similar concerns, optimizing operations for a mixed

Table 1 Overview of references on the role of LFVs and ECBs in sustainable LM delivery

References	Modeling approach[a]	Fixed fleet	Het. vehicles	Electric cargo bikes	Problem[b]	Det. versuss stoch.	Solution methods		
							E	H	Type[c]
Ćirović et al. (2014)	OM	✓	✓	–	GVRP	D	–	✓	SA
Schneider et al. (2014)	OM	✓	–	–	VRP	D	–	✓	VNS, TS
Goeke and Schneider (2015)	OM	✓	✓	–	HVRP	D	–	✓	ALNS
Choubassi et al. (2016)	OM	✓	–	✓	VRP	D	–	✓	RRP
Hiermann et al. (2016)	OM	–	✓	–	FSMVRP	D	✓	✓	B&P, ALNS
Anderluh et al. (2017)	OM	✓	✓	–	2eVRP	D	–	✓	GRASP
Melo and Baptista (2017)	SM	–	✓	✓	MTS	–	–	–	–
Fikar et al. (2018)	SM	–	✓	✓	ABS	–	–	–	–

(continued)

Table 1 (continued)

References	Modeling approach[a]	Fixed fleet	Het. vehicles	Electric cargo bikes	Problem[b]	Det. versuss stoch.	Solution methods		
							E	H	Type[c]
Zhang et al. (2018)	SM	–	✓	✓	MTS	–	–	–	–
Hiermann et al. (2019)	OM	✓	✓	–	HVRP	D	–	✓	GA, LLNS
Jie et al. (2019)	OM	✓	✓	–	2eVRP	D	✓	✓	CG, ALNS
Rezgui et al. (2019)	OM	–	✓	–	FSMVRP	D	–	✓	VNS
Dalla Chiara et al. (2020)	SM	–	✓	✓	PS	–	–	–	–
Enthoven et al. (2020)	OM	✓	✓	–	2eVRP	D	–	✓	ALNS
Anderluh et al. (2021)	OM	✓	✓	✓	2eVRP	D	–	✓	LNS
Caggiani et al. (2021)	OM	✓	✓	✓	2eVRP	D	✓	–	MILP
Al-dal'ain and Celebi (2021)	OM	–	✓	–	FSMVRP	D	✓	–	RM
Llorca and Moeckel (2021)	SM	–	✓	✓	UFFS	–	–	–	–
Mühlbauer and Fontaine (2021)	OM	✓	✓	–	2eVRP	D	–	✓	PLNS

(continued)

Table 1 (continued)

References	Modeling approach[a]	Fixed fleet	Het. vehicles	Electric cargo bikes	Problem[b]	Det. versuss stoch.	Solution methods		
							E	H	Type[c]
Fontaine (2022)	OM	✓	–	✓	VRP	D	–	✓	ALNS
Malladi et al. (2022)	OM	–	✓	–	FSMVRP	S	–	✓	SAA, ALNS
Wang and Zhao (2023)	OM	–	✓	–	FSMVRP	D	–	✓	LNS
Achamrah and Puchinger (2024)	OM	✓	✓	–	2eVRP	S	–	✓	ADAM
Beatrici et al. (2025)	OM	–	✓	✓	FSMVRP	S	✓	✓	MILP, KSBH

[a] *OM:* Optimization Model, *SM:* Simulation Model

[b] *GVRP:* Green VRP, *HVRP:* Heterogeneous VRP, *FSMVRP:* Fleet Size and Mix VRP, *2eVRP:* Two-echelon VRP, *MTS:* Microscopic Traffic Simulation, *ABS:* Agent-Based Simulation, *PS:* Performance Simulation, UFFS: Urban Freight Flows Simulation

[c] *SA:* Simulated Annealing, *VNS:* Variable Neighborhood Search, *TS:* Tabu Search, *ALNS:* Adaptive Large Neighborhood Search, *RRP:* Ruin and Recreate Principle, *B&P:* Branch & Price, *GRASP:* Greedy Randomize Adaptive Search Procedure, *GA:* Genetic Algorithm, *LLNS:* Local Large Neighborhood Search, *CG:* Column Generation, *LNS:* Large Neighborhood Search, *MILP:* Mixed Integer Linear Program, *RM*: Replacement Model, *PLNS:* Parallelized Large Neighborhood Search, *SAA:* Sample Average Approximation, *ADAM:* ADAptive Moment estimation-based algorithm, *KSBH:* Kernel Search Based Heuristic

The *"Modeling Approach"* column specifies whether the study employs an Optimization Model (OM) or a Simulation Model (SM). The next three columns (*"Fixed Fleet"*, *"Het. Vehicles"*, and *"Electric Cargo Bikes"*) provide information about the characteristics of the vehicle fleet considered. The *"Problem"* column describes the type of routing problem addressed, while *"Det. vs Stoch."* indicates whether uncertainty is explicitly modeled. Finally, the last three columns report the solution method used: exact (E) and/or heuristic (H) and the specific algorithm or technique applied to solve the proposed model fleet under time window constraints. Hiermann et al. (2019) further expand on this by considering a heterogeneous fleet consisting of conventional, plug-in hybrid, and electric vehicles, while also accounting for vehicle recharging operations.

The majority of these contributions adopt a deterministic framework. However, uncertainty, such as fluctuating demand or variable travel times, can undermine the reliability of such solutions. Among the reviewed works, only Achamrah and Puchinger (2024) explicitly incorporate stochasticity, proposing a 2eVRP with uncertain customer demand in a setting involving transshipment operations and time windows.

Solution Methods As shown in Table 1, heuristic approaches dominate due to the computational intractability of exact methods in large-scale VRPs. For example, Jie et al. (2019) implement a hybrid method combining column generation (CG) with Adaptive Large Neighborhood Search (ALNS), while Anderluh et al. (2017) use a Greedy Randomize Adaptive Search Procedure (GRASP) combined with path relinking.

ALNS is also applied in Goeke and Schneider (2015), enhanced with local search, and in Enthoven et al. (2020), where specialized destroy operators are tailored to the hierarchical nature of the 2eVRP. Additional local search heuristics include Variable Neighborhood Search (VNS), (see Schneider et al. 2014), Large Neighborhood Search (LNS), (see Anderluh et al. 2021), Parallelized Large Neighborhood Search (PLNS) (see Mühlbauer and Fontaine 2021), and Local Large Neighborhood Search (LLNS), (see Hiermann et al. 2019). All of these approaches rely on problem-specific destroy-and-repair mechanisms.

Ćirović et al. (2014) implement Simulated Annealing (SA) to train an adaptive neural network that supports routing decisions. Finally, for the stochastic model in Achamrah and Puchinger (2024), a gradient-descent-based heuristic is developed, where the ADAM optimization algorithm is employed.

3.2 Fleet Size and Mix Problems in Sustainable LM Delivery

Several studies have investigated how fleet sizing and composition influence the efficiency and environmental performance of LM delivery services and are predominantly addressed under a deterministic framework.

One of the early contributions is Hiermann et al. (2016), who propose an FSMVRP involving electric vehicles with time window constraints. Their model also incorporates decisions about charging station placement and vehicle recharging operations. Similar considerations are addressed by Wang and Zhao (2023), who propose a partial linear recharging strategy within a comparable FSMVRP framework. Rezgui et al. (2019) also tackle a time-constrained FSMVRP, introducing modular electric components for LFVs as a sustainable alternative in urban delivery. In Al-dal'ain and Celebi (2021), electric technologies are explored further: an FSMVRP is formulated using a replacement model to assess optimal planning for mixed fleets composed of CVs and electric LFVs.

While deterministic formulations remain the most common, stochastic optimization offers a powerful framework to account for real-world uncertainties. Nonetheless, only a few studies have adopted this approach. Notably, Malladi et al. (2022) address an FSMVRP with uncertain delivery requests and customer demand. In their model, fleet size and composition are determined tactically, while operational decisions are made without full knowledge of future demand. To manage the risk of unserved requests, the model includes the option to outsource deliveries to external providers.

Solution Methods Similar to the fixed-fleet problems discussed in Sect. 3.1, FSMVRPs are predominantly solved using heuristic methods. For example, VNS is applied in Rezgui et al. (2019), while Wang and Zhao (2023) adopt LNS to solve their FSMVRP under time window and recharging constraints.

Several hybrid heuristics have also been developed. Hiermann et al. (2016) combine ALNS with an exact Branch-and-Price (B&P) method, while Malladi et al. (2022) integrate ALNS with a Sample Average Approximation (SAA) technique to address demand uncertainty.

Only a limited number of exact methods have been proposed. Al-dal'ain and Celebi (2021) introduce a replacement-based MILP model to evaluate different fleet configurations and routing plans, solving small-to-moderate instances exactly.

3.3 *Electric Cargo Bikes in LM Delivery*

Among Light Freight Vehicles, ECBs represent a particularly promising solution for enhancing the sustainability of LM logistics, as demonstrated by a growing body of literature summarized in Table 1.

A number of simulation-based studies have investigated the potential of ECBs to improve delivery efficiency while mitigating the negative externalities of urban freight transport. Melo and Baptista (2017) evaluate the impact of ECBs on urban mobility, environmental quality, and overall delivery performance. Fikar et al. (2018) emphasize the importance of fleet sizing, showing that a sufficient number of ECBs can ensure timely and sustainable deliveries. Zhang et al. (2018) highlight the economic and environmental benefits of ECB adoption, especially in terms of reduced delivery costs and improved quality of life. Llorca and Moeckel (2021) further demonstrate that ECBs can contribute significantly to the reduction of air and noise pollution in dense urban areas. Dalla Chiara et al. (2020) also explore these benefits, focusing in particular on how ECBs can reduce parking time and improve delivery punctuality.

While simulation studies underline the strategic value of ECBs, only a limited number of contributions propose optimization models that directly incorporate ECBs into LM delivery planning. Choubassi et al. (2016) investigate the substitution of conventional vehicles with cargo cycles in a real-world VRP for mail distribution. Anderluh et al. (2021) develop a multi-objective 2eVRP with vehicle synchronization to address accessibility constraints in dense urban centers, explicitly considering emissions and noise alongside economic criteria. Caggiani et al. (2021) propose a 2eVRP model that includes time windows, partial recharging operations, and ECB-based deliveries. Fontaine (2022) addresses the effect of payload-dependent travel times for ECBs, showing that neglecting load-dependency can lead to time window violations in urban routes.

As with the broader class of LM delivery problems, most studies involving ECBs adopt a deterministic framework. As for stochastic optimization approaches, indeed, to the best of our knowledge only in Beatrici et al. (2025) propose a stochastic FSMVRP with consistent routing, in which routes are designed during the tactical phase and then adjusted in the operational phase based on the realization of uncertain customer demand. Their real-world case study demonstrates that including ECBs in the fleet increases the flexibility and adaptability of routing strategies under uncertainty.

Solution Methods In line with findings from Sects. 3.1 and 3.2, heuristic methods dominate in solving ECB-related LM optimization problems. Heuristics such as LNS (Anderluh et al. 2021), ALNS (Fontaine 2022), and the Ruin and Recreate Principle (RRP), (see Choubassi et al. 2016) are commonly employed to handle the complexity of these problems.

Exact methods remain relatively rare, but notable examples include the MILP proposed by Caggiani et al. (2021) to solve small 2eVRP instances. Beatrici et al. (2025) also propose an exact solution method for synthetic instances, and develop a path-based reformulation combined with a Kernel Search Based Heuristic (KSBH) to address larger, real-world cases.

4 Conclusions and Future Research

Sustainable LM logistics is an increasingly important area of research in Operations Research, particularly due to the need to reduce the environmental and social impacts of freight transport in urban areas. In this context, LFVs and, more specifically, ECBs, represent valuable alternatives to conventional vehicles, offering a practical means to reduce emissions, alleviate congestion, and improve the quality of life in cities.

This review has shown that LFVs and ECBs can ensure service reliability while contributing to a cleaner and more accessible urban environment. Their operational flexibility makes them particularly suited for navigating dense urban centers and responding to the constraints of modern e-commerce-driven logistics.

A number of future research directions emerge:

- *Optimizing the deployment of ECBs*: while simulation studies have demonstrated the benefits of ECBs in urban logistics, few optimization models explicitly incorporate their operational constraints, such as limited range, payload capacity, and recharging requirements. Future research should develop integrated optimization frameworks that accurately model these aspects and assess how ECBs can be effectively deployed within multi-vehicle, multi-echelon delivery systems.
- *Improving exact methods*: heuristic and metaheuristic methods dominate the solution landscape due to the complexity of the problems involved. However, more attention should be given to exact and hybrid approaches, particularly for small- and medium- scale instances where accurate solutions are required, such as in policy design or infrastructure planning.
- *Incorporating uncertainty*: although the relevance of uncertainty in urban logistics is widely recognized, relatively few studies address it explicitly. Future work should expand the use of stochastic, robust, and distributionally robust optimization to model uncertainty in demand, travel time, energy consumption, and other key parameters affecting light mobility (see Maggioni et al. 2017; Perboli et al. 2017; Filippi et al. 2025).
- *Dynamic and data-driven approaches*: real-time data availability from urban IoT infrastructures and connected vehicles offers the opportunity to move from static to dynamic routing models, where decisions are continuously adapted based on live data. This direction, combined with machine learning techniques for demand forecasting and parameter estimation, could significantly improve the responsiveness of LM systems using light vehicles.
- *System-level integration and policy implications*: light mobility should not be viewed in isolation, but as part of a multimodal urban transport ecosystem. Future research could explore the integration of ECBs with public transport, parcel lockers, and urban consolidation centers. Moreover, models should increasingly support policy design, helping municipalities evaluate regulatory incentives, infrastructure investments, and zoning restrictions to foster the adoption of sustainable delivery modes.

In summary, LM offers tangible opportunities to achieve more sustainable and resilient urban logistics. Bridging the current modeling gaps, particularly in the treatment of uncertainty, dynamic decision-making, and policy-oriented design, will be essential to unlocking the full potential of LFVs and ECBs in future smart cities.

Acknowledgements This study was carried out within the MOST (Sustainable Mobility National Research Center) and received funding from the European Union Next-GenerationEU (PIANO NAZIONALE DI RIPRESA E RESILIENZA (PNRR) MISSIONE 4 COMPONENTE 2, INVESTIMENTO 1.4 D.D. 1033 17/06/2022, CN00000023), Spoke 5 "Light Vehicle and Active Mobility". This manuscript reflects only the authors' views and opinions, neither the European Union nor the European Commission can be considered responsible for them.

References

Achamrah FE, Puchinger J (2024) A gradient-descent-based framework for solving a stochastic two-echelon delivery problem with cargo-bikes. Transp Res Part E Logist Transp Rev 189:103677

Al-dal'ain R, Celebi D (2021) Planning a mixed fleet of electric and conventional vehicles for urban freight with routing and replacement considerations. Sustain Cities Soc 73:103105

Anderluh A, Hemmelmayr VC, Nolz PC (2017) Synchronizing vans and cargo bikes in a city distribution network. CEJOR 25:345–376

Anderluh A, Nolz PC, Hemmelmayr VC et al (2021) Multi-objective optimization of a two-echelon vehicle routing problem with vehicle synchronization and 'grey zone' customers arising in urban logistics. Eur J Oper Res 289(3):940–958

Beatrici P, Birolini S, Maggioni F, Malighetti, P (2025) Stochastic fleet size and mix consistent vehicle routing problem for last mile delivery. arXiv preprint arXiv:2512.09764

Bezzi D, Jabali O, Maggioni F (2023) A threshold recourse policy for the electric vehicle routing problem with stochastic energy consumption. In: International conference on optimization and decision science. Springer, Cham, pp 219–229

Birge JR, Louveaux F (2011) Introduction to stochastic programming. Springer Science & Business Media.

Blazejewski L, Sherriff G, Davies N (2020). Delivering the last mile: scoping the potential for E-cargo bikes

Caggiani L, Colovic A, Prencipe LP et al (2021) A green logistics solution for last-mile deliveries considering e-vans and e-cargo bikes. Transp Res Procedia 52:75–82

Choubassi C, Seedah DPK, Jiang N et al (2016) Economic analysis of cargo cycles for urban mail delivery. Transp Res Rec 2547(1):102–110

Ćirović G, Pamučar D, Božanić D (2014) Green logistic vehicle routing problem: routing light delivery vehicles in urban areas using a neuro-fuzzy model. Expert Syst Appl 41(9):4245–4258

Dalla Chiara G, Alho AR, Cheng C et al (2020) Exploring benefits of cargo-cycles versus trucks for urban parcel delivery under different demand scenarios. Transp Res Rec 2674(5):553–562

Enthoven D, Jargalsaikhan B, Roodbergen KJ et al (2020) The two-echelon vehicle routing problem with covering options: city logistics with cargo bikes and parcel lockers. Comput Oper Res 118:104919

Filippi C, Maggioni F, Speranza MG (2025) The robust shortest path: a review and open problems. Comput Oper Res 182:107096

Fikar C, Hirsch P, Gronalt M (2018) A decision support system to investigate dynamic last-mile distribution facilitating cargo-bikes. Int J Log Res Appl 21(3):300–317

Fontaine P (2022) The vehicle routing problem with load-dependent travel times for cargo bicycles. Eur J Oper Res 300(3):1005–1016

Goeke D, Schneider M (2015) Routing a mixed fleet of electric and conventional vehicles. Eur J Oper Res 245(1):81–99

Hiermann G, Puchinger J, Ropke S et al (2016) The electric fleet size and mix vehicle routing problem with time windows and recharging stations. Eur J Oper Res 252(3):995–1018

Hiermann G, Hartl RF, Puchinger J et al (2019) Routing a mix of conventional, plug-in hybrid, and electric vehicles. Eur J Oper Res 272(1):235–248

Jie W, Yang J, Zhang M et al (2019) The two-echelon capacitated electric vehicle routing problem with battery swapping stations: formulation and efficient methodology. Eur J Oper Res 272(3):879–904

King AJ, Wallace SW (2012) Modeling with stochastic programming. Springer

Koç Ç, Bektaş T, Jabali O et al (2016) Thirty years of heterogeneous vehicle routing. Eur J Oper Res 249(1):1–21

Llorca C, Moeckel R (2021) Assessment of the potential of cargo bikes and electrification for last-mile parcel delivery by means of simulation of urban freight flows. Eur Transp Res Rev 13(1):33

Maggioni F, Potra F, Bertocchi M (2017) A scenario-based framework for supply planning under uncertainty: stochastic programming versus robust optimization approaches. Comput Manag Sci 14(5):5–44

Malladi S, Christensen JM, Ramirez D et al (2022) Stochastic fleet mix optimization: evaluating electromobility in urban logistics. Transp Res Part E Logist Transp Rev 158:102554

Melo S, Baptista P (2017) Evaluating the impacts of using cargo cycles on urban logistics: Integrating traffic, environmental and operational boundaries. Eur Transp Res Rev 9:1–10

Moghdani R, Salimifard K, Demir E et al (2021) The green vehicle routing problem: a systematic literature review. J Clean Prod 279:123691

Mohammad WAM, Nazih DY, Elomri A et al (2023) Innovative solutions in last mile delivery: concepts, practices, challenges, and future directions. Supply Chain Forum: Int J 24:151–169

Mühlbauer F, Fontaine P (2021) A parallelised large neighbourhood search heuristic for the asymmetric two-echelon vehicle routing problem with swap containers for cargo-bicycles. Eur J Oper Res 289(2):742–757

Nocerino R, Colorni A, Lia F et al (2016) E-bikes and E-scooters for smart logistics: environmental and economic sustainability in pro-E-bike Italian pilots. Transp Res Procedia 14:2362–2371

Perboli G, Tadei R, Vigo D (2011) The two-echelon capacitated vehicle routing problem: models and math-based heuristics. Transp Sci 45(3):364–380

Perboli G, Gobbato L, Maggioni F (2017) A progressive hedging method for the multi-path travelling salesman problem with stochastic travel times. IMA J Manag Math 28(1):65–86

Rezgui D, Siala JC, Aggoune-Mtalaa W et al (2019) Application of a variable neighborhood search algorithm to a fleet size and mix vehicle routing problem with electric modular vehicles. Comput Ind Eng 130:537–550

Shahmohammadi S, Steinmann ZJN, Tambjerg L et al (2020) Comparative greenhouse gas footprinting of online versus traditional shopping for fast-moving consumer goods: a stochastic approach. Environ Sci Technol 54(6):3499–3509

Schneider M, Stenger A, Goeke D (2014) The electric vehicle-routing problem with time windows and recharging stations. Transp Sci 48(4):500–520

Spinelli A, Bezzi D, Jabali O, Maggioni F (2026) A stochastic electric vehicle routing problem under uncertain energy consumption. Transp Res Part C: Emerg Technol 183:105480

Toth P, Vigo D (2014) Vehicle routing: problems, methods, and applications. SIAM

Wang W, Zhao J (2023) Partial linear recharging strategy for the electric fleet size and mix vehicle routing problem with time windows and recharging stations. Eur J Oper Res 308(2):929–948

Wrighton S, Reiter K (2016) CycleLogistics—Moving Europe forward! Transp Res Procedia 12:950–958

Zhang L, Matteis T, Thaller C et al (2018) Simulation-based assessment of cargo bicycle and pick-up point in urban parcel delivery. Procedia Comput Sci 130:18–25

Designing Slow Tourism with EVSE Technologies: A Critical Review

Rossella Moscarelli

Abstract The chapter discusses a collection of case studies on cycle paths incorporating EVSE technologies. The aim is to analyse the installation process of EVSE services in different European contexts, focusing on the location, goals and implementation process of the projects, as well as the private or public actors involved. The analysis tackles solely with cases related to extra-urban and leisure mobility, thus excluding urban and commuter uses. It represents one of the first syntheses on this topic and demonstrates the importance of private actors in developing these projects, as well as the random distribution of EVSE services: e-bikes services are still seen exclusively as products to sell, with no precise plan to integrate them with slow travel infrastructures and the territory.

1 Designing Slow Tourism with e-services

This paper reflects on the relatively new and original field of e-slow tourism by presenting an analysis of existing routes equipped with e-bike recharging services. Seven cases have been selected in Europe and nine in Italy. This study aims to understand the criteria and approaches that have guided the location of e-bikes charging stations (CSs).

The analysis thus addresses cycling tourism by considering its relationship with the so-called e-velomobility, which concerns "*practices, systems and technologies of e-cycling, specifically around electrically-assisted bicycles, where a combination of pedal power and (optional) electrical assistance propels the rider*" (Behrendt 2018: 5). As discussed in this volume (see chapter 1.5 by Pirolo and Pileri) and in many other studies (Han et al. 2020; Senese et al. 2023a, b; Lagerstedt and Svensson 2022; Pröbstl-Haider et al. 2018), e-technology can produce a positive effect on slow tourism by increasing the number of potential users. Thanks to e-bikes, people can cycle for long periods without specific training, even in mountainous and hilly areas (Scuttari and Isetti 2019).

R. Moscarelli (✉)
Department of Architecture and Urban Studies, Politecnico Di Milano, Milano, Italy
e-mail: Rossella.moscarelli@polimi.it

© The Author(s), under exclusive license to Springer Nature Switzerland AG 2026
R. Moscarelli et al. (eds.), *The Future of Slow Mobility*, Research for Development,
https://doi.org/10.1007/978-3-032-23660-9_12

In order to promote the use of e-bikes in slow tourism practices, specific facilities are required, such as a suitable charging infrastructure. The location of e-bike CSs depends firstly on economic factors, such as the available budget, which affects the number of stations, and secondly on technological issues related to e-bike batteries, which can vary depending on the type of vehicle and the route's topography (the steeper the slope, the shorter the battery life) (Burani et al. 2022). Moreover, the location of e-bike CSs along a cycle tourist path can also be viewed as a more complex spatial decision based not only on economic and technical constraints. In fact, it can contribute to defining breaks during the slow tourism experience (Pirolo and Moscarelli 2025; see chapter 2.6 by Pirolo, Moscarelli and Belotti). Where to stop along a cycle path is usually unpredictable, but many attractive factors, such as cultural heritage, accommodation services or transport hubs, play a crucial role (Ma 2018; Lee 2014). As a new service, the presence of e-bike CSs can also influence travellers to stop in a specific place. Where should we plan for slow travellers to stop and recharge their e-bikes? Should it be directly along the cycle path, or in urban areas where, while they recharge the bike, they can also have a coffee in a bar, visit a cultural attraction, or even spend the night? Should local businesses, such as restaurants and hospitality services, be involved in the maintenance and management of e-bike CSs? Who should promote the development of e-services for slow tourism, and what should the goals and ambitions be? These are just a few of the questions related to the issue of CS locations. Different answers will have different impacts on the territory and the traveller's experience.

The chapter discusses an analysis that attempts to determine which design principles were used to plan EVSE services along existing cycle paths.

The paper is structured as follows. First, the cases are presented alongside the analysis methodology. Then, one sample is described in more detail to demonstrate the approach used to analyse all the other cases. Finally, a synoptic view of the results obtained is introduced and discussed.

2 The Case Studies and the Methodology of Analysis

The analysis is presented as a series of case studies, some of which are located in Europe (identified by the code 'En') and some of which are located in Italy (identified by the code 'In') (Fig. 1). These are referred to e-bike cycle paths.

The list of European cases include:

- Main cycleway (E1), in Germany. It is a path of more than 500 kms along the Main River (it mostly runs along its banks), which goes from Mainz to Bischofsgrün (538 km) or Creußen (557 km);
- Danube cycleway (E2), German section, spanning 600 kms along the national border. It starts in the Black Forest area where the Danube's source river, the Breg, originates and, then, continues along the Danube's banks until Passau;

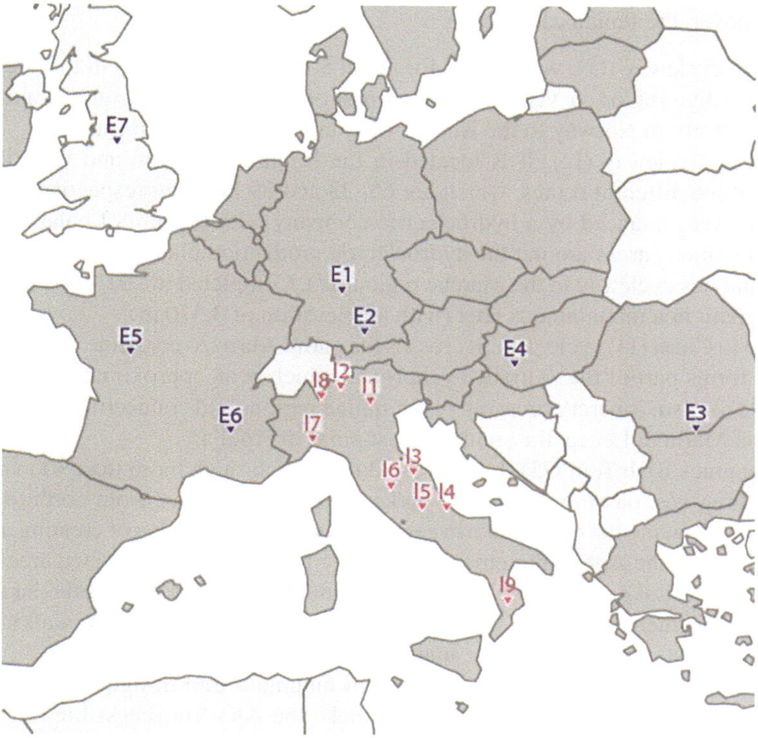

Fig. 1 Distribution of the European and Italian case studies. *Source* Developed by the author

- Danube Cycleway (E3), Romanian and Bulgarian sections, which is the final part of the EuroVelo 6 cycle route connecting the Atlantic Ocean in France to the Black Sea in Romania. The path at the Romanian-Bulgarian border is yet to be realised, though an e-bike charging station network has already been established;
- Balaton Lake cycleway (E4), in Hungary. It is a 237 km circular route around the lake and it is part of the EuroVelo 14, which traverses Austria and Hungary and establishes a crucial east–west corridor;
- *La Loire à Vélo* (E5), in France. It is one of the country's most important cycle routes, stretching from Saint-Nazaire to Nevers over 900 kms. It represents the starting point for EuroVelo 6;
- *La Bella* Via (E6), in France, located in the Auvergne-Rhône-Alpes region. It runs from Valence to Chanaz or La Balme-de-Sillingy via a 225 or 280 km path;
- National Cycle Network (E7), in United Kingdom. Four paths in the network that already have CSs have been selected: the Devon Coast to Coast route (159 km) and the Cotswold Line route (69 km) in England, the National Route 6 route (711 km) and the Hadrian's Cycleway route (England, 278 km) and Caledonia Way (Scotland, 374 km).

Moreover, the Italian cases are:

- Adige cycleway (I1), which runs for about 80 km from Merano in the Trentino-Alto Adige region to Verona. It forms part of the EuroVelo 7 route, connecting Cape North in Norway to the island of Malta in the Mediterranean;
- Energy Cycleway (I2). It is located in the Lombardy region and it is divided into three different routes, which are 65, 23 and 39 km long respectively. These paths were financed by a hydroelectric company with the aim of enhancing the mountainous areas around the hydroelectric production plants;
- Apennines cycleway in the Marche region (I3). Completed in 2023, it is a 198 km ring path in a mountainous area (with an elevation of 3,570 m);
- Bike to Coast (I4) cycle path. It crosses the entire Abruzzo coastline over 131 kms and forms part of the Adriatic cycle route, which is an approximately 1,300-km cycle tourism itinerary crossing seven Italian regions and connecting Trieste with Santa Maria di Leuca, the southernmost point of Apulia;
- Apennines Bike Tour (I5). It is a 3,500 km route through more than 300 villages in 14 regions, passing through 56 parks and protected areas, from north to south Italy. Following the earthquake in central Italy in 2016, the idea of creating a cycle route along the Apennines emerged, initially focusing on the affected areas;
- Trasimeno Lake cycleway (I6). It is a 58-km circular path around the lake. The Lake Trasimeno Hoteliers' Union has promoted other 25 different walking and cycling routes connected to the main circular route;
- Alta Via dei Monti Liguri (I7), a 440 km mountain trail designed for walking, crossing Liguria from west to east. To make the Alta Via accessible to cycling tourists, variants of the original route have been identified and free e-bike recharging services are available at many accommodation facilities along the route. The route is not suitable for inexperienced cyclists, as it is a mountain trail with a frequently bumpy and stony surface and significant altitude differences;
- Cycle paths in Stelvio National Park (I8), suitable for mountain biking and follow silvopastoral and old military roads dating back to the Great War. E-bike recharging stations have been placed along the routes to extend the experience to those with less experience, enabling them to explore the mountain park with the help of an electric bike;
- Cycleway of Calabrian Parks (I9), which runs for 545 km along the Calabrian Apennines, alongside low-traffic roads, tracks and cycle paths. The project is based on the promotion of regional nature areas and their settlements. The parks connected by this route are Pollino National Park, Sila National Park, Serre Regional Nature Park and Aspromonte National Park.

The cases were selected by considering the main relevant cycling paths and their distribution. Information on all the cases was collected through direct contact with local stakeholders (via email or telephone interviews), as well as from websites and publications. References for each case can be found in the bibliography.

All cases were analysed using a similar approach. First data collected refers to the total number of CSs and the average distance between them along the path. Then, the focus is on the characteristics of the CSs and where they are located, e.g. directly along the cycling path, in urban areas close to it, or in specific places such as public areas (e.g. squares), tourism facilities (e.g. hotels) or amenities (e.g. museums). These aspects are highly relevant, as they suggest a potential correlation between the provision of this e-bike service and the territory and local activities. The final part of the analysis delves into the installation process of EVSE services (the reasons, aims, and manner in which the project was realised) and the actors involved (private or public). This section is unique to this analysis, as it is unusual to observe how e-mobility is integrated into regional and urban planning processes, as mentioned in the topic overview.

It should be noted that the structure of the analysis remains consistent, even when not all the required information could be found or all aspects could be explored in depth.

3 A Sample Analysis: e-bike CSs Along the National Cycle Network in the UK

The installation of e-bike CSs along the National Cycle Network (NCN) in the UK has been selected as an example to illustrate the approach used for the other cases more clearly.

It is useful to mention that the NCN covers every region and nation of the United Kingdom, with a total length of over 25,000 km. Routes include quiet roads, traffic-free paths and greenways. The network was first conceived in 1997 by the sustainable transport charity Sustrans, which remains its custodian, although the Network is a joint production. In fact, the routes are provided and maintained by local governments, private landowners, large and small, volunteers, and government and non-governmental bodies.

Figure 2 shows the actual distribution of e-bike charging stations along the five cycle routes included in the NCN: the Devon Coast to Coast (England), the Cotswold Line (England), National Route 6 (England), Hadrian's Cycleway (England) and the Caledonia Way (Scotland).

The distribution of CSs along cycling paths varies considerably depending on their length and the number of CSs. In particular, we can count:

- two CSs along the 159 km Devon—Coast to Coast route, with an average charging station distance of almost 80 km;
- one CS along the 69 km Cotswold Line;
- two CSs along National Route 6 (711 km long), with an average distance between charging stations of over 355 km;
- one CS along the Hadrian's Cycleway (218 km long);
- three CSs along the 374 km Caledonia Way, with an average distance between charging stations of almost 125 km.

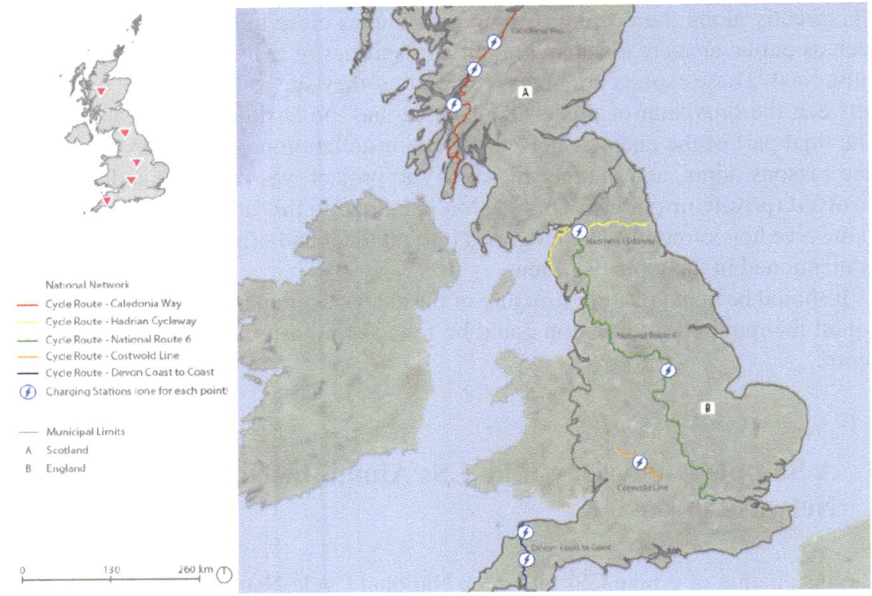

Fig. 2 Distribution of the e-bike CSs along the National Netwotk (UK). Source: developed by the author

Once we have identified the total number of CSs and their distribution along the cycling paths, we can consider their location. Firstly, we observe whether they are close to the route, i.e. within a distance range of 0–100 m. Secondly, we consider their relation to specific services or facilities connected with tourism, culture or social activities. Tables 1 and 2 summarise the results obtained for the NCN case.

Finally, the last part of the analysis looks at how the cycle paths were provided with EVSE services and how these services were managed.

In this case, the two main partners involved were Sustrans and Bosch. Sustrans is a UK-based charity dedicated to enhancing walking, wheeling and cycling as sustainable modes of transport. It was founded in 1977 and is the promoter of the initiative. For this project, it has partnered with Bosch to install e-bike charging stations along well-known National Cycle Network routes. The initiative is intended not only to encourage the adoption of electric bikes and improve tourist services, but

Table 1 Location (along the cycleway or not) of the e-bike CSs along the National Netwotk (UK)

CS along the cycleway [#]	CS not along the cycleway [#]			
0–100 m	100–500 m	500–1000 m	1000–2500 m	> 2500 m
2	2	1	1	2

Source Developed by the author

Table 2 Location (in relation to the activities) of the e-bike CSs along the National Netwotk (UK)

Location of e-bike CS	CS along the cycleway (100–500 m) [#]	CS not along the cycleway (> 500 m) [#]	Total [#]
Accomodation facilities	1	–	1
Commercial areas	–	4	4
Cultural heritage	–	1	1
Visitor center	1	1	2
Total [#]	**2**	**6**	**8**

Source Developed by the author

also to promote climate policies and sustainable transport, making it more accessible and convenient for individuals to explore the country by bike.

The CSs are located indoors and are equipped with charging cables. Cycle parking is located nearby. Each station has six charging compartments with lockable doors for security, allowing simultaneous use. To charge an e-bike, simply detach the battery, connect it to the charging cable inside a compartment and lock it with the provided key.

These charging stations are free to use for all e-bike riders and currently support Bosch batteries exclusively. However, future phases of the project will expand compatibility to include other electric bike battery systems.

4 Collecting the Results: The European and the Italian Cases

A similar analysis has been conducted for all 16 cases. Here, we present a collection of the results for comparison and to obtain a synthetic overview.

Starting with the number and distribution of CSs along the different cycling paths, it is possible to observe quite different results (Table 3). Even if this aspect should consider the maximum battery range (that correspond to 50kms in plain conditions), we have cases with a huge amount of charging stations (such as the Bike to Cost-I4 case with an average distance of charging stations of 6 km) and others with a less than a minimum number (such as the La Loire à Vélo-E5 with an average distance of charging stations of 75 km).

We can then consider whether the e-bike CSs are located close to the route (i.e. within a distance range of 0–100 m) or not. The data for each case are shown in Table 4. The aggregation of the European cases results shows that CSs are mainly located relatively far from the cycleway. This means that tourists need to deviate from the route to reach the EVSE service and explore the local area. Conversely, the Italian cases reveal an opposite approach, with EVSE services mainly located along the cycle path, i.e. within 100 m. This could be interpreted as less strategic

Table 3 Distrubution of the e-bike CSs

Cycleway	Length	CS [#]	Average distance of CS
E1	a. Mainz to Bischofsgrün: 538 km b. Mainz to Creußen: 557 km	a. 133 b. 124	a. 4 km b. 4.5 km
E2	598 km	92	6,5 km
E3*	n.d	34	n.d
E4	237 km	42	5.6 km
E5	900 km	12	75 km
E6	a. Valence to Annecy: 280 km b. Valence to Chambery: 225 km	a. 1 b. 1	a. 140 km b. –
E7	a. Devon—Coast to coast: 159 km b. Cotswold line: 69 km c. National route 6: 711 km d. Hadrian's cycleway: 218.1 km e. Caledonia way: 374 km	a. 2 b. 1 c. 2 d. 1 e. 3	a. 79.5 km b. – c. 355.2 km d. – e. 124.6 km
I1	80 km	25	3.2 km
I2	127 km	4	31.7 km
I3	198 km	11	18 km
I4	131 km	22	5.9 km
I5	3,500 km	44	79.5 km
I6	124 km	30	4.1 km
I7	440 km	42	10.5 km
I8*	n.d	21	n.d
I9*	n.d	n.d	n.d

Source Developed by the author
* Cases of cycleways which are still under development and construction

positioning of the charging stations, which are strictly related to the path but less connected to places that tourists could discover by stopping to charge their e-bikes.

In more detail, we can consider the location of e-bike CSs in relation to their urban surroundings. Of the total number of e-bike CSs analysed (#521), we can identify the following:

- 140 located along streets, of which 86 in European cases and 54 in Italian cases;
- 126 located in public areas such as squares, gardens and urban parks, of which 76 in European cases and 50 in Italian cases);
- 106 are close to accommodation facilities (huts, campsites, and hotels), of which 35 are in European cases and 71 in Italian cases;
- 70 are close to commercial areas (restaurants, bars, and shops), of which 56 are in European cases and 14 are in Italian cases;
- 22 are close to sports and education facilities (sports fields, gymnasiums, swimming pools, schools and kindergartens), of which 14 are in European cases and 8 in Italian cases;

Table 4 Location (along the cycleway or not) of the e-bike CSs

Cycleway	CS along the cycleway [#]	CS not along the cycleway [#]			
	0–100 m	100–500 m	500–1000 m	1000–2500 m	> 2500 m
E1	28	32	27	29	17
E2	29	20	10	11	22
E3	1	2	–	–	31
E4	19	10	2	4	7
E5	7	1	2	1	1
E6	1	–	–	–	–
E7	2	2	1	1	2
Total E—Cases	**87**	**67**	**42**	**46**	**80**
I1	1	–	4	4	16
I2	2	1	–	–	1
I3	11	–	–	–	–
I4	7	3	2	5	5
I5	36	7	1	–	–
I6	14	15	1	–	–
I7	17	5	7	6	7
I8	14	1	–	6	–
I9	n.d	n.d	n.d	n.d	n.d
Total I—Cases	**102**	**32**	**15**	**21**	**29**

Source Developed by the author

- 21 are close to cultural heritage and facilities (monuments, artefacts, historical sites, museums, libraries and visitor centres), of which 19 are in European cases and two in Italian cases;
- 9 are close to transport hubs (train stations), all of which are in European cases;
- 5 are close to entertainment facilities (festival halls, water parks, zoos and recreation centres), all in European cases.

The remaining 21 cases are located close to other activities (not strictly related to tourism or recreational places such as municipality building, NGO or cemetery).

Moving to the analysis of the process, from each case we can deduce the reasons, directly or implicitly declared, to install the charging stations (Table 5). We see that they are mainly related to improving tourist services. This is the only aspect common to both the European and Italian cases. On the contrary the other reasons appear quite different. In European cases the issues of climate protection, sustainable transportation and accessibility are cited. While in Italian ones, the EVSE services are seen also as an occasion to slowly experience hilly and mountain places and to revitalize, though slow tourism, marginal areas.

Moreover, if we consider the promoter, namely who has activated the process of installing charging stations, and the sponsor, namely who has financed it, we see (Table 6) how the private and public actors are both involved, in European and Italian cases. This seems to suggest that the issue of EVSE services need a cooperation of private stakeholders (usually the companies that produce or distribute the charging stations) and public entities, from local to over local ones. However, it is necessary to remember that the private actor is much faster than the public one in realizing its project, as emerged from the analysis of the promoters. This aspect needs to be carefully considered since it can produce a over distribution of EVSE services, seen exclusively as products to sell, without a precise plan to integrate them with slow travel infrastructures and with the territory.

5 Some Preliminary Reflections on the Design of e-bike CSs

This paper presents a series of case studies on tourist cycleways incorporating EVSE technologies, i.e. e-bike charging stations. It aims to analyse the installation process of EVSE services in various European settings, focusing on the location, rationale and objectives of the projects, and the stakeholders involved, whether private or public. This analysis is one of the first syntheses of this topic and could provide valuable insights for the design of future projects.

A key finding is the variety of cases, which are not all related to cycleways with similar technical characteristics (e.g. infrastructure type or length). Some of them have well-structured projects, such as the Danube and UK cycleways, while others are still in the early stages of development, such as the Italian Apennines Bike Tour. This suggests that electric charging stations are sometimes being built before cycling infrastructure is put in place. Moreover, we can see that the number of charging stations varies greatly from case to case. This issue probably arises due to a weak plan for EVSE services, which could be located depending on the budget or other factors, rather than being implemented effectively alongside the cycleway. From this perspective, we can also see that the decision to improve the cycle paths with e-bike CSs is sometimes made before the infrastructure itself is realised. This can be a critical issue if it reveals that EVSE services are regarded as more of a marketing tool than a way of improving the quality of existing paths. Alternatively, as the analysis of the process revealed, the installation of e-bike services is encouraged by private companies (some of which are described in the promoters' section) or by contingent public funds that are not part of a broader strategic framework on slow mobility.

A similar variety can be found in the locations of CSs. In Europe, they are mainly located far from cycleways, along local streets and in public and commercial areas. It is also relevant that they are located in relation to municipal buildings and train stations. Conversely, Italian cases reveal an opposite approach, with EVSE services mainly located within 100 m of cycle paths. This could be interpreted as less strategic

Table 5 Main reasons for installing e-bike CSs

Cycleway	Reason						
	Improving touristic services	Climate protection	Enhancing sustainable transport	Improving accessibility	Energy efficiency	Regenerating area (e.g. earthquake area)	Increasing attractiveness of mountains/hills
E1	X	X					
E2	X	X					
E3				X			
E4	X	X	X	X			
E5	X						
E6	X						
E7	X	X	X				
I1	X						
I2					X		
I3	X					X	
I4	X						
I5	X					X	
I6	X						
I7	X						X
I8	X						X
I9						X	

Source Developed by the author

Table 6 Analysis of the promoters and sponsors of e-bike CSs

Cycleway	Promoter sponsor		
	Private	Public	Description
E1	X	X	Promoters: • Franken Tourismus and Local Transportation Associations (for 109 charging stations) • Local Action Groups (LAGs) and municipalities which are in "Wald erFahren project" (for 24 charging stations) Sponsors: • Public and private entities for 109 charging stations • European Agricultural Fund for Rural Development (EAFRD) for 24 charging stations
E2	X		Promoters and Sponsors: ADFC (General German Bicycle Club), Germany Top River Routes, and WGD Donau Oberösterreich Tourismus GmbH (Danube Upper Austria Tourism Association)
E3	X	X	Promoters and Sponsors: • Agency for Regional Development and Business Center—Vidin (Bulgaria), Vidin chamber of commerce and industry, Romanian Association for Electronic Industry and Software Oltenia, Local Employers' Association for Small and Middle Enterprises Calafat • Interreg Europe
E4		X	Promoters and Sponsors: • Interreg Europe • The Lake Balaton Development Coordination Agency
E5	X	X	Promoters: • Région Pays de la Loire • Région Centre Val de Loire • Comité Régional du Tourisme Centre—Val de Loire • Solutions & co Sponsors: • Région Pays de la Loire • Région Centre Val de Loire • European Regional Development Fund(ERDF)
E6	X	X	Promoters: • Auvergne Rhône-Alpes Region, the Department of Isère, and the Pays Voironnais • Tourisme Pays Voironnais Sponsors: • Auvergne Rhône-Alpes Region, the Department of Isère, and the Pays Voironnais
E7	X		Promoters and Sponsors: • Sustrans and Bosch
I1	X	X	Promoters and Sponsors: • Municipalities • Private Entities
I2	X		Promoters and Sponsors: • A2A hydroelectric company

(continued)

Table 6 (continued)

Cycleway	Promoter sponsor		
	Private	Public	Description
I3		X	Promoters: • Municipalities • Mountain Union of Catria and Nerone Sponsors: • SNAI Regione Marche • Municipalities
I4	X	X	Promoters and Sponsors: • Municipalities • Private Entities
I5	X	X	Promoters: • Local Operators • Tour Operators linked with territory • Municipalities Sponsors: Municipalities
I6	X		Promoters and Sponsors: • DINAclub—Repower
I7	n.d	n.d	n.d
I8	X	X	Promoters and Sponsors: For Lombardy Region: • Regione Lombardia-Direzione Sport e Politiche per i giovani • Interreg Italy-Switzerland (ERDF + Switzerland Fund) For Autonomous Province of South Tyrol and the Autonomous Province of Trento: • Specific agreement with accommodation facilities
I9		X	Promoters and Sponsors: • Department of Environmental Protection of the Region of Calabria • Regional Parks and Protected Natural Areas sector

Source Developed by the author

localisation of CSs, which are strictly related to the path, but less connected to places that slow tourists could discover by stopping to charge their e-bikes.

Turning to the analysis of the process, we can deduce the reasons, whether directly or implicitly declared, for installing charging stations from each case. We see that they are mainly related to improving tourist services. This is the only aspect common to both the European and Italian cases. Conversely, the other reasons are quite different. In the European cases, climate protection, sustainable transportation and accessibility are cited as issues. In Italian cases, EVSE services are also seen as an opportunity to experience hilly and mountainous areas slowly and to revitalise marginal areas through slow tourism.

In summary, the analysis demonstrated the importance of private actors in developing e-bike CSs, as well as the random distribution of such services. The general picture shows that e-bike services are primarily viewed as products to sell, with no specific plans in place to integrate them with slow travel infrastructures and the territory.

Finally, we can conclude by highlighting the main limitations of the analysis. Firstly, the variety of cases discussed above makes them difficult to compare; they should be considered in their particularities rather than in a comparative way. Furthermore, the analysis could be improved by enriching the procedural aspects. This analysis provides a starting point from which to gain a broad overview and decide which cases to investigate in more depth.

Acknowledgements The paper is a result of the study carried out within the MOST—Sustainable Mobility National Research Center and received funding from the European Union Next-Generation EU (PIANO NAZIONALE DI RIPRESA E RESILIENZA (PNRR)—MISSIONE 4 COMPONENTE 2, INVESTIMENTO 1.4—D.D. 1033 17/06/2022, CN00000023). This chapter reflects only the authors' views and opinions; neither the European Union nor the European Commission can be considered responsible for them.

Many thanks to Eleonora Santoro and Nicola Petaccia for their contribution to the analysis.

References

Behrendt F (2018) Why cycling matters for electric mobility: towards diverse, active and sustainable e-mobilities. Mobilities 13:64–80

Burani E, Cabri G, Leoncini M (2022) An algorithm to predict e-bike power consumption based on planned routes. Electronics 11:1105

Han H et al (2020) Cycling tourism: a perspective article. Tourism Rev 75:162–164

Lagerstedt E, Svensson H (2022) Do tourists dream of electric bikes?: Electric bikes as a mean to improve sustainability of tourism in rural Sweden. In: Proceedings of the 25th international academic Mindtrek conference, pp 167–178

Lee CF (2014) An investigation of factors determining cycling experience and frequency. Tour Geogr 16(5):844–862

Ma Z (2018) Tourists' preference of the long-distance cycling tourism based on content analysis. Sch. J. Econ. Bus. Manag. 5(3):167–173

Pirolo L, Moscarelli R (2025) Integrating cycling tourism, e-mobility and local development: a spatial model to plan slow tourist paths for E-bikes. Reg Sci Policy Pract 17:100237

Pröbstl-Haider U, Lund-Durlacher D, Antonschmidt H, Hödl C (2018) Mountain bike tourism in Austria and the Alpine region–towards a sustainable model for multi-stakeholder product development. J Sustain Tour 26:567–582

Senese A et al (2023a) The role of e-bike in discovering geodiversity and geoheritage. Sustainability 15:4979

Scuttari A, Isetti G (2019) E-mobility and sustainable tourism transport in remote areas—Insights from the Alpine case study of South Tyrol (IT). Zeitschrift Für Tourismuswissenschaft, De Gruyter 11(2):237–256

Case Studies References

E1:

https://www.mainradweg.com/rund-ums-rad/akku-ladestationen/karte/#map-list. Accessed 12 Apr 2024

https://www.mainradweg.com/streckenfuehrung/kurzinfo/. Accessed 17 Apr 2024

https://walderfahren.de/. Accessed 17 Apr 2024

E2:

Meschik M (2021) The success of the cycle tourist backbone along the Danube in Germany and Austria. In: Pileri P, Moscarelli R (eds) Cycling & walking for regional development. How slowness regenerates marginal areas. Springer, Switzerland, pp 235–249

Pfaffenbichler P et al (2021) Planning and designing infrastructure and services for sustainable bicycle tourism along the EuroVelo routes in the Danube region. Road and rail infrastructure, conference proceedings 6th international conference on road and rail infrastructure, pp 895–901.

https://www.deutsche-donau.com/danube-cycle-path/service/e-bike-recharging/. Accessed 29 Apr 2024

https://www.adfc.de/ueber-uns#c924. Accessed 29 Apr 2024

https://www.donauregion.at/service/b2b-bereich.html. Accessed 29 Apr 2024

E3:

https://www.google.com/maps/d/viewer?mid=1LtchABatoLWT_Q5fhGMSJkurEvWiHYKA&femb=1&ll=43.79523192582661%2C25.57646339999998&z=7. Accessed 27 Mar 2024

https://www.theplanetedit.com/danube-cycle-path-guide/. Accessed 27 Mar 2024

https://www.interregrobg.eu/en/e-bike-net.html. Accessed 27 Mar 2024

https://keep.eu/projects/17916/E-bike-Net-EN/. Accessed 27 Mar 2024

https://www.interregeurope.eu/good-practices/e-bike-net. Accessed 27 Mar 2024

E4:

Lőrincz K, Banász Z, Csapó J (2020) Customer involvement in sustainable tourism planning at Lake Balaton, Hungary—analysis of the consumer preferences of the active cycling tourists. Sustainability 12(12):5174

https://projects2014-2020.interregeurope.eu/fileadmin/user_upload/tx_tevprojects/library/file_1554388396.pdf. Accessed 06 Apr 2024

https://projects2014-2020.interregeurope.eu/fileadmin/user_upload/tx_tevprojects/library/file_1613559418.pdf. Accessed 06 Apr 2024

https://keep.eu/projects/18835/Delivering-Efficient-Sustai-EN/. Accessed 06 Apr 2024

E5:

Bres R et al. (2023) Analysis of cycling network evolution in OpenStreetMap through a data quality prism. AGILE: GIScience Series 4:1–9

https://www.loirebybike.co.uk/nos-bonnes-idees/bike-services-on-the-loire-a-velo/. Accessed 27 Mar 2024.

https://www.loirebybike.co.uk/homepage/la-loire-a-velo-nature-culture-and-adventure/. Accessed 27 Mar 2024

https://www.francevelotourisme.com/itineraire/la-loire-a-velo-eurovelo-6. Accessed 20 Mar 2024

E6:

https://www.isere-tourisme.com/equipements/veloroute-la-belle-en-pays-voironnais. Accessed 20 Mar 2024

https://www.isere-tourisme.com/equipements/la-belle-de-la-balme-de-sillingy-ou-de-chanaz-valence-haute-savoie-savoie-isere-drome. Accessed 20 Mar 2024

https://www.francevelotourisme.com/itineraire/la-belle-via. Accessed 20 Mar 2024

https://www.isere-tourisme.com/commerces-et-services/aire-de-service-cyclo-belle-vallee-de-lisere. Accessed 27 Mar 2024

E7:
Brice X (2021) The United Kingdom's national cycle network: paths for everyone, past, present and future. In: Pileri P, Moscarelli R (eds) Cycling & walking for regional development. How slowness regenerates marginal areas. Springer, Switzerland, pp 205–219
Gouais AL et al (2021) A natural experimental study of new walking and cycling infrastructure across the United Kingdom: the Connect2 programme. J Transp Health 20:100968
https://www.sustrans.org.uk/our-blog/national-cycle-network-route-collections/2023/all/find-an-e-bike-charging-station-on-the-national-cycle-network/. Accessed 29 Apr 2024
https://www.sustrans.org.uk/find-other-routes/cotswold-line/. Accessed 29 Apr 2024
https://www.sustrans.org.uk/find-other-routes/the-caledonia-way/. Accessed 29 Apr 2024
https://www.sustrans.org.uk/find-a-route-on-the-national-cycle-network/route-6/. Accessed 29 Apr 2024
https://www.sustrans.org.uk/find-other-routes/hadrians-cycleway/. Accessed 29 Apr 2024
I1:
https://www.visitrovereto.it/vivi/outdoor/in-sella/bike/stazioni-ricarica-e-bike/. Accessed 24 Jan 2024
https://www.cr-ager.it/chi-siamo/. Accessed 24 Jan 2024
https://www.comune.rovereto.tn.it/Servizi/Ricaricare-la-bicicletta-elettrica. Accessed 24 Jan 2024
https://www.ladige.it/territori/vallagarina/2021/08/18/sul-monte-baldo-arrivano-le-colonnine-di-ricarica-per-le-bici-elettriche-1.2976874. Accessed 24 Jan 2024.
https://www.crvallagarina.it/cassa-green/i-nostri-interventi-green/colonnine-ricarica-ebike-e-auto/. Accessed 24 Jan 2024.
https://www.giornaletrentino.it/montagna/vallarsa-amica-delle-bici-elettriche-attiva-una-nuova-stazione-di-ricarica-1.3238469. Accessed 24 Jan 2024
I2:
https://www.cicloviadellenergia.eu/#contenitore_mappa_principale. Accessed 24 Jan 2024
https://www.bormio.eu/it/ciclovia-dell-energia-percorso-rosso. Accessed 24 Jan 2024
I3:
https://visitaltemarche.it/ciclo-appeninica/. Accessed 12 Apr 2024
https://visitaltemarche.it/download/caam-generale/. Accessed 12 Apr 2024
https://www.regione.marche.it/Entra-in-Regione/Fondi-Europei/Progetti-realizzati/Post/97629/Ciclovia-delle-Alte-Marche. Accessed 12 Apr 2024
https://comune.cagli.ps.it/inaugurazione-ciclo-appenninica-alte-marche-caam-sabato-22-aprile-2023-sassoferrato/. Accessed 12 Apr 2024
https://www.fieradelcicloturismo.it/expo/ciclovia-appenninica-delle-alte-marche/ (12.04.2024)
Interview by mail/call with Luigi Soriani (Technical Instructor in Mountain Union of Catria and Nerone). Accessed 12 Apr 2024
I4:
Bianchi A, D'Uva D, Rolando A (2020) An innovational digital tool in Gis procedure: mapping Adriatic coast in Abruzzo region to support design of slow mobility routes. The International Archives of the Photogrammetry, Remote Sensing and Spatial Information Sciences/ International Archives of the Photogrammetry, Remote Sensing and Spatial Information Sciences, XLIII-B4–2020, p. 533–537.
https://cicloturismo.abruzzoturismo.it/. Accessed 24 Apr 2024
https://abruzzoturismo.it/itinerari/bike-coast-da-martinsicuro-san-salvo-lungo-la-verde-delladriatico. Accessed 24 Apr 2024
https://abruzzoturismo.it/it/cicloturismo-0. Accessed 24 Apr 2024
https://www.regione.abruzzo.it/content/turismo-importante-garantire-servizi-inclusivi-ai-disabili#:~:text=Nello%20specifico%20il%20progetto%20riguarda,%C3%A8%20in%20grado%20di%20offrire. Accessed 24 Apr 2024
https://www.fieradelcicloturismo.it/2024/04/03/abruzzo-in-bicicletta-gli-itinerari-e-gli-eventi-2024/. Accessed 24 Apr 2024
https://www.fieradelcicloturismo.it/expo/abruzzo/. Accessed 24 Apr 2024

https://fiabitalia.it/bike-to-coast-abruzzo/. Accessed 24 Apr 2024
https://www.fiab.info/download/biketocoast.pdf. Accessed 24 Apr 2024
https://umap.openstreetmap.fr/it/map/ciclovia-adriatica_916184#9/42.0394/14.3329. Accessed 24 Apr 2024
https://www.frentanabike.it/servizi-per-i-ciclisti/. Accessed 24 Apr 2024
https://www.vaielettrico.it/la-costa-dei-trabocchi-svolta-grazie-alle-bike-il-nostro-viaggio/. Accessed 24 Apr 2024
https://costadeitrabocchimob.it/consigli/noleggia-una-bici-19433. Accessed 24 Apr 2024
https://www.ricariche-bike.it/ricariche-bike-installazioni-stazioni-ricarica-biciclette-elettriche-in-tutta-italia/. Accessed 24 Apr 2024
https://www.comune.pescara.it/node/319#:~:text=Per%20soli%200%2C50%20%E2%82%AC,pescara.ecospazio.it%. Accessed 24 Apr 2024
I5:
https://appenninobiketour.com/. Accessed 24 Apr 2024
https://www.fieradelcicloturismo.it/expo/legambiente/. Accessed 24 Apr 2024
I6:
https://www.lagotrasimeno.net/it/bicicletta-c29. Accessed 24 Apr 2024
https://www.lagotrasimeno.net/it/colonnine-elettriche-pubbliche-c39. Accessed 24 Apr 2024
https://www.lagotrasimeno.net/it/repower-s-p-a-a159. Accessed 30 Apr 2024
I7:
https://www.ebikeliguria.it/. Accessed 24 Jan 2024
https://www.altaviainfoh24.com/. Accessed 24 Jan 2024
I8:
Senese A et al (2023b) The role of E-Bike in discovering geodiversity and geoheritage. Sustainability 15(6):4979
https://www.bormiobike.it/it/stazioni-ricarica-ebike-alta-valtellina/. Accessed 24 Jan 2024
https://www.e-stelvio.it/parco-nazionale-dello-stelvio.html. Accessed 24 Jan 2024
https://www.suedtirol.info/it/it/esperienze-eventi/plp-esperienze/o/stazioni-di-ricarica. Accessed 24 Jan 2024
https://www.outdooractive.com/it/poi/parco-nazionale-dello-stelvio/punto-di-ricarica-e-bike/27638923/. Accessed 24 Jan 2024
https://www.calendariovaltellinese.com/blog/colonnine-ebike-vallespluga. Accessed 24 Jan 2024
I9:
https://www.cicloviaparchicalabria.it/it/organizza-il-viaggio/aree-ricarica-e-bike. Accessed 31 Jan 2024
https://inviaggio.touringclub.it/news/nasce-la-ciclovia-dei-parchi-la-calabria-sceglie-la-bicicletta-per-ripartire. Accessed 31 Jan 2024
https://www.cicloturismo360.it/cicloturismo-in-calabria-nasce-la-ciclovia-dei-parchi/3292/. Accessed 31 Jan 2024

Planning Slow Tourism with e-Technologies. A Proposal for Locating e-Bike Charging Stations Along the Italian National System of Tourist Cycling Routes

Luca Pirolo, Rossella Moscarelli, and Pietro Belotti

Abstract This chapter explores the potential integration of Electric Vehicle Supply Equipment (EVSE) technologies with slow tourism. An existing model developed by the authors to define the location of e-bike charging stations is applied to the most relevant Italian planning strategies related to cycle tourism, i.e., the National System of Tourist Cycling Routes. This study enables us to refine certain aspects of the model, such as its application to cycle routes crossing areas with different orography, and demonstrate the usability and scalability of this optimization tool.

1 Planning Slow Tourism Together with EVSE Technologies Through Optimization Models

This chapter explores the joint planning of e-technologies within a slow tourism project. We focus on integrating Electric Vehicle Supply Equipment (EVSE) services along slow tourist routes, primarily addressing the issue of deploying suitable charging infrastructure. In the case of a cycleway, this means that e-bike charging stations (CSs) must be strategically located to ensure comprehensive coverage of the entire route. The location of e-bike CSs has been studied less than that of e-car CSs, although some work on the topic can be found. This has mainly been approached as a technical issue concerning battery autonomy (Micari et al. 2017; Belotti et al. 2024).

L. Pirolo (✉) · P. Belotti
Department of Electronics, Information, and Bioengineering—Politecnico diMilano, Milano, Italy
e-mail: luca.pirolo@polimi.it

P. Belotti
e-mail: pietro.belotti@polimi.it

R. Moscarelli
Department of Architecture and Urban Studies—Politecnico di Milano, Milano, Italy
e-mail: rossella.moscarelli@polimi.it

© The Author(s), under exclusive license to Springer Nature Switzerland AG 2026

R. Moscarelli et al. (eds.), *The Future of Slow Mobility*, Research for Development,
https://doi.org/10.1007/978-3-032-23660-9_13

Consequently, the location of CSs depends solely on the requirement to distribute an adequate number of EVSE services at fixed intervals along the path.

Recent studies by the authors (Pirolo and Moscarelli 2025; Pirolo et al. 2024) have discussed the development of a mathematical model that considers other criteria for planning the location of e-bike CSs, while also considering a possible local development strategy. This practical tool is an optimization model implemented in a GIS (Geographic Information System) environment that can support the planning decision about the location of e-bike CSs.

The use of mathematical models in tourism planning has been extensively studied over the decades, with diverse case studies focusing on the unique characteristics of such journeys. In fact, the Tourist Trip Design Problem (TTDP) (Van Steenwegen and Van Oudheusden 2007) is a fundamental optimization problem in tourism trip planning. It aims to create personalised itineraries that mimic travellers' decision-making processes while enhancing their experiences. Addressing the complexities of planning, such as the numerous options and constraints, the TTDP seeks to find the best possible solutions for visiting points of interest (POIs) while taking into account limits on distance, budget, time, and environmental conditions. Some scholars have applied this typological problem to the specific issue of cycle tourism (Černá et al. 2014; Verbeeck et al. 2014), by introducing the so-called Cycle Trip Planning Problem (CTPP).

Based on this background, our study proposes an optimization model integrating key elements of the TTDP framework with technical constraints (e.g., budget and distance depending on battery autonomy) and spatial criteria. These spatial criteria, processed by the model, prioritise locations in marginal areas based on the assumption that the presence of charging stations (CSs) would encourage e-cyclists to stop in 'undiscovered' places. The availability of charging stations could encourage slow travellers to stop at certain places and rest for a few hours (e.g., to eat or visit attractions) or longer (e.g., an overnight stay). In this case, the location of CSs would contribute to determining the locations where breaks would be taken during the journey, and thus the economic impact of the slow tourism project.

The model enables us to link technological improvements in EVSE to the concept of regenerating marginal territories. Stopping to charge an e-bike can encourage people to explore a place, providing an opportunity for economic and sustainable development in areas that are not popular with tourists. This approach aligns with the idea that slow tourism can stimulate economic growth in remote areas such as small rural or mountain villages, which are usually excluded from tourist routes (Gazzola et al. 2018). It should be noted that slow tourism can generate significant demand for services related to the consumption of local food and craft products, accommodation (e.g., hotels and bed and breakfasts), enjoyment of places (e.g., museums and cultural experiences), and other specific services (e.g., equipment repairs and technical apparel) (Petroman et al. 2021; Pileri and Moscarelli 2021). Considering these positive externalities, we can redefine the issue of locating e-bike charging stations (CSs) across the territory, rather than considering it as a blind kilometer-based distribution. The problem of finding the optimal locations for e-bike CSs therefore arises, based on spatial criteria and integrating the usual technical constraints regarding budget and battery autonomy.

This chapter demonstrates the application of an optimization model to the extensive Italian National System of Tourist Cycling Routes (hereby, SNCT, using the Italian acronym).

2 The Case Study: The Italian National System of Tourist Cycling Routes

The SNCT was launched in 2016 by the Italian Ministry of Transport, in collaboration with the Ministry of Culture, to:

- identify priority axes, namely the cycle tourism routes, of the Italian tourist cycle path;
- allocate national funds dedicated to the construction of cycle paths included in the plan;
- define a project management mechanism for the construction of each cycle route included in the plan.

In 2016, the following four cycle routes were initially included: VENTO, SOLE, GRAB, and the Apulian Aqueduct Cycle Path. In 2017, a further six were added. Today, the SNCT therefore includes ten cycle routes totaling over 7000 km (Pileri and Moscarelli 2021).

Seven of the SNCT cycle routes[1] have been chosen as case studies (Fig. 1, Table 1) on which to experiment with the mathematical model for defining EVSE location for three main reasons. First, they are still a work in progress, since some of the routes already exist, while the rest will be completed in the coming years through national and regional funding. Moreover, there is currently no planning strategy in place for EVSE locations along cycleways. Finally, they cross marginal and inner areas comprising small rural villages, where it is worthwhile to associate the technological improvements of the EVSE with a regenerative idea.

The following paragraphs will consider how the optimization model was applied to the seven selected cycle routes by discussing the model's underlying methodology and the main results obtained.

[1] The GRAB route has been excluded from the analysis due to its urban nature and confinement within the administrative boundaries of Rome. The Garda cycleway is excluded due to the controversial environmental issues surrounding the project. The Vento cycleway is excluded as it has already been examined in Pirolo and Moscarelli (2025).

Fig. 1 The seven cases studied selected from the Italian National System of Tourist Cycling Routes. (1) Trieste-Lignano Sabbiadoro-Venice Cycle Route; (2) Sole Cycle; (3) Tyrrhenian Cycle Route; (4) Adriatic Cycle Route; (5) Apulian Aqueduct Cycle Route; (6) Magna Grecia Cycle Route; (7) Sardinian Cycle Route. *Source* Elaboration by authors based on Moscarelli

3 The Methods: Elaborating a Model to Locate e-Bike CSs

The method used and described in this section was first developed by Pirolo and Moscarelli (2025). A detailed explanation of the criteria and the underlying reasoning can be found in that work. In this contribution, and specifically in the following paragraphs, several improvements to the original model are introduced. These enhancements include modifications that allow the methodology to be applied to different

Table 1 Cycleways considered for the study, with their length and the number of municipalities crossed by the infrastructure

Cycleway	Length (km)	Crossed municipality (#)
Trieste-Lignano Sabbiadoro-Venice	150	30
Sole	300	73
Tyrrhenian	870	97
Adriatic	820	78
Apulian Aqueduct	500	62
Magna Grecia	1000	85
Sardinian	1230	110
Total	4870	535

Source Developed by the authors

orographic contexts, each characterized by distinct landscape and morphological features.

The optimization model is a mathematical tool that enables the representation and solution of complex decision-making problems using equations and inequalities. In this study, it is applied to the design of a cycling tourism network.

The first step in solving the model is to build the network, defining its characteristics by selecting nodes and arcs. This process was carried out using GIS software to accurately represent the real scenario. The nodes represent the points, namely the urban areas where cyclists can stop, park their bikes, eat a meal, visit a place of interest, or stay overnight in accommodation. The arcs, on the other hand, correspond to the sections of the cycling or road network that connect the different nodes.

As described in the study objective, identifying nodes with specific characteristics is essential for obtaining reliable results from the mathematical model. The characterization of nodes must therefore follow a methodical and consistent process.

To promote a regenerative effect in the areas crossed by the cycling path, the charging infrastructure for electric bicycles must be integrated harmoniously into the urban context of small and medium-sized towns. In the Italian context, the heart of urban settlements can generally be identified as the town square, often located near the town hall or a church.

The squares considered eligible for selection by the model are not located directly on the main cycling path, but they are located at a maximum distance of 2500 m from it. Bicycle charging Stations (CSs) can be located in these spaces, allowing cyclists to easily reach the main recreational areas and places of local interest by walking. The square is therefore defined as a geographical reference point to which a network node is assigned.

The overall objective of the study is to use cycling tourism as a tool for widespread territorial regeneration, capable of triggering sustainable development processes and strengthening synergies with existing public policies, such as the National Strategy for Inner Areas (SNAI). The underlying idea is that the creation of a well-structured cycling tourism network is not only an infrastructure for slow mobility, but also a driver for the enhancement of small towns and marginal areas, promoting new economic, social, and tourism opportunities.

In order to identify the areas where the intervention can generate the greatest positive impact, an evaluation system based on the classification of urban centers has been developed.

The methodology adopted considers three main factors, chosen for their ability to represent the vulnerability and regenerative potential of the territories, as discussed in the next paragraphs.

3.1 The Marginality Score to Assess the POIs

The first factor is the inclusion of the municipality in the national database of Inner Areas (UVAL 2014), which identifies territories distant from large urban centers and recipients of specific support policies. The second concerns the demographic dimension, measured by the number of inhabitants, a useful indicator for understanding the scale of the local context. Finally, the third factor considers the variation in the youth population, a parameter that reflects the phenomena of depopulation and the loss of attractiveness of smaller towns. Each of these three indicators was discretized according to the methods described in Table 2 and translated into a numerical value.

The sum of the scores obtained for the three factors produces a summary index called the *Marginality Score*, calculated as follows:

$$Marginality\ Score = IA + Pop + YPL$$

where *IA* represents the indicator of belonging to Inner Areas, *Pop* the score relating to the resident population, and *YPL* the loss of young population.

Table 2 Definition of marginality score

Marginality score	Score	
Inner area (*IA*)	Yes	5
	No	0
Population size (*Pop*)	< 5000	8
	between 5000 and 10,000	5
	> 10,000	2
Youth population loss (*YPL*)	Degrowth over 10%	5
	Degrowth within 10%	2
	Positive trend	0

Source Developed by the authors

This score enables a comparative assessment of the level of marginality across the different territories traversed by the cycling tourism network. In this way, the optimization model can favor, among the various possible destinations, those that offer the greatest potential for socio-economic regeneration, thus contributing to a more balanced and sustainable territorial development process.

3.2 *Identifying Spatial Criteria and Scoring the POIs*

In addition to structural and territorial aspects, each node in the network is assigned a specific score, which represents its ability to generate value for the surrounding area. This score, described in detail below, is based mainly on the number and type of services available within the urban settlement considered.

To determine this score, a circular buffer of 500 m has been defined around the main square and the route connecting it to the main cycling path. The area thus delimited allows the services available within the proximity to be identified and counted, including the entire route that the cyclist takes to reach the urban center.

The combination of the connecting route, the square, and the buffer zone constitutes a Point of Interest (POI), a key element in evaluating the network. The POIs are divided into three categories: 'Very marginal' (category A) when the score is above 12; 'medium marginal' (category B) when the score is between 6 and 12; and 'not marginal' (category C) when the score is under 6.

The score of the POI, and therefore of the node, is calculated as the weighted sum of various contributions, each of which expresses an aspect considered strategic for the initiation of widespread urban regeneration processes in the context analyzed. Formally, the formula can be expressed as follows:

$$OI_{score} = \sum_{i=1}^{N} w_i k_i = \sum \alpha k_1 + \beta k_2 + \gamma k_3 + \cdots$$

where k_i represents each parameter considered and w_i the weight assigned to it, which is a value between 0 and 1.

The main parameters used to define the score are as follows:

1. the distance of the square from the main cycling path;
2. the safety of the route connecting the cycling path to the square, assessed based on the type and quality of the road;
3. the presence of essential services (e.g. hotels, pharmacies, or shops) that support the local community and attract visitors;
4. the availability of food services, such as bars and restaurants, which encourage cycling tourists to stay;
5. the presence of staffed cultural services (e.g. museums, information centers, or exhibition spaces) with dedicated personnel;
6. the presence of cultural heritage within a radius of 250 m from the POI boundary;

7. the preference for less attractive urban contexts, to favor areas that are currently less valued but have high regenerative potential. The *Marginality Score* is taken into account here.

A higher score indicates that the POI is more suitable for selection within the model's optimal solution, highlighting its strategic importance in the sustainable territorial development process. Table 3 summarizes the value of each criteria:

Table 3 Definition of the criteria

Criteria	Quantitative measurement	Score (pt)	
1. Distance from the cycleway	Road distance from detour (m)	1–500 m	5
		500–1000 m	4
		1000–1500 m	3
		1500–2000 m	2
		2000–2500 m	1
2. Road safety level	Primary road category calculated as a percentage of the total length of the branch (%)	Main suburban roads	1
		Secondary extra-urban roads	2
		Main urban roads	3
		Secondary urban roads	4
		Local roads	6
		Cyclable urban streets	7
		Cycle-pedestrian paths	7
3. Presence of essential services	Binary value	Absent	0
		At least one present	1
4. Presence of refreshment services	Binary value	Absent	0
		At least one present	1
5. Presence of cultural services (e.g., museums)	Binary value	Absent	0
		At least one present	1
6. Presence of cultural heritage	Binary value	Absent	0
		At least one present	1
7. Marginality score	Calculation of the phase 2	Non-marginal	1
		Medium marginal	2
		Very marginal	3

Source Developed by the authors

3.3 *Introducing Different Battery Constraints Depending on the Morphology of Places*

Another fundamental element in building up the model concerns the definition of the battery range of electric bicycles, a parameter closely linked to the morphological characteristics of the territory crossed. Since range varies according to slope and elevation, it was necessary to adopt a representative value for each geographical context.

To simplify the problem, it was decided to assign different range values depending on the average orography of the cycling path, referring to the technical performance of electric bicycle models currently on the market. The average elevation gain was calculated as the ratio between the total positive elevation gain and the total length of the route. In practice, only the positive elevation gains are added up, and the value obtained is divided by the total length of the route. This measure is also commonly used in cycling statistics, as it provides a concise but effective indicator of the degree of difficulty of the route.

$$gain_{avg} = \frac{\sum_{i=0}^{n-1} \max(0,\ h_{i+1} - h_i)}{\sum_{i=0}^{n-1} d_i}$$

where h_i represents the POI altitude and d_i is the distance between two consecutive POIs. Three representative scenarios were defined:

- Plain: average gradient less than 0.20;
- Hill: average gradient between 0.21 and 0.40;
- Mountain: average gradient greater than 0.41.

Each cycling route belonging to the SNCT has been associated with one of these scenarios, together with a corresponding level of electric assistance. To determine the actual range of electric bicycles, the BOSCH propulsion and power system (Bosch eBike Systems 2025) was used as a reference. Its official website allows users to configure the technical specifications of an e-bike and estimate its range based on the selected parameters.

The reference values used for the simulation are as follows:

- Propulsion system: Performance Line PX
- Battery: CompactTube 500
- Average speed: 18 km/h
- Total weight (cyclist + bicycle): 120 kg
- Pedaling frequency: 60 rpm.

The BOSCH system allows you to select up to seven levels of assistance, but to simplify the application of the model, the number has been reduced to three. Each level is associated with a different estimated range, as shown below:

- ECO → 122 km

- TOUR → 88 km
- TURBO → 70 km.

After calculating the average elevation gain of each cycle route, each route was associated with one of the three orographic scenarios defined above.

Each scenario was then linked to a level of assistance consistent with the difficulty of the route:

- Plain scenario → ECO level
- Hill scenario → TOUR level
- Mountain scenario → TURBO level.

This distinction allows the model to be calibrated realistically, taking into account the different physical conditions of the routes and the resulting energy consumption of electric bicycles. In this way, the optimization model integrates not only territorial and infrastructural variables but also technical aspects related to cycling tourism, ensuring a coherent and functional representation of the system as a whole.

3.4 The Final Optimization Model

Consider a graph $G(N, A)$, where N represents the set of the nodes equal to the number of POI considered in the cycling tourist network, and A is the set of arcs that connect the nodes. Let us also define a set of paths $k = \{p_1, p_2, \ldots, p_k\}$, in which each path is defined as a sequence of nodes $p_h = \left(i_1^h, i_2^h, \ldots, i_{qh}^h\right)$, where q_h indicates the number of nodes belonging to the path p_h, with $i_r^h \in N, \forall h \in [k]$. The selected paths correspond to the longest paths obtainable in the network, such that they reach at least 80% of the total nodes. For each arc $(i, j) \in A$ is associated with a distance d_{ij}, while the sum of the distances between two consecutive POIs where charging stations are installed is indicated by l_{ij} for each path p. For each node $j \in N$ the cost of installing the charger is defined c_j, and the total installation costs cannot exceed the available budget b. The battery life of an electric bicycle is represented by the parameter δ. For further information on the model used for a single cycle path, see the specifications in Pirolo and Moscarelli (2025).

The objective of the model is to maximize the sum of the scores σ associated with POIs where a charging station is installed, according to the technical and budgetary constraints.

The mathematical model is formally described as follows:

$$max_{z,x} \sum_{(i,j)\in A} \sigma_j x_{ij} \tag{1}$$

$$s.t. \quad l_{(ij)}^h x_{(ij)}^h \leq \delta \quad \forall (i,j) \in A_h, \forall p \in P \tag{2}$$

$$\sum_{j \in N} c_j z_j \leq b \tag{3}$$

$$\sum_{j \in N:(i,j) \in A} x_{ij}^h - \sum_{j \in N:(j,i) \in A} x_{j,i} = \begin{cases} -1 & \textit{if } i = \omega_h \\ 1 & \textit{if } i = \alpha_h \\ 0 & \textit{otherwise} \end{cases} \quad \forall i \in N_h, \forall h \in [k] \tag{4}$$

$$z_j = \sum_{(i,j) \in BS_h(j)} x_{ij}^h \quad \forall j \in p_h, \forall h \in [k] \tag{5}$$

$$x_{ij}^h \in \{0, 1\} \quad \forall (i, j) \in A_h, \forall h \in [k] \tag{6}$$

$$z_j \in \{0, 1\} \quad \forall j \in V \tag{7}$$

The binary variables x_{ij}^h and z_j define respectively:

- $x_{ij}^h = 1$ if arc (i, j) is used in path h;
- $z_j = 1$ if a CS is installed in node j.

Constraint 2 ensures that the distance between two consecutive charging stations does not exceed the battery range δ, expressed in kilometers. Constraint 3 represents the budget limit, requiring that the total cost of the installations remain below the available threshold. Constraint 4 guarantees the existence of a valid route from α_h to ω_h for each path h, where α_h and ω_h indicate the first and last unknown nodes of the route, respectively. Finally, constraint 5 links the variables z and x, representing the so-called *linking constraint*, which ensures consistency between the selection of nodes and the arcs actually used.

4 The Results of the Analysis: The e-Bike CSs Along the Italian SNCT

The mathematical model developed was applied to cycling paths belonging to the Italian SNCT to identify the optimal configuration for the positioning of CSs for recharging e-bikes. The application of the model produced results of considerable interest, both from a quantitative point of view and in terms of territorial and strategic implications. The main results obtained are now illustrated and discussed, following the same methodological logic presented in the previous section.

4.1 Assigning POIs on Marginality

In the SNCT case study, 571 POIs were identified within a radius of 2500 m from the seven main cycling routes.

The POIs were classified using the Marginality Score, introduced into the model as a summary indicator of the degree of territorial marginality.

Of the 571 POIs identified:

- 268 (47% of the total) are *very marginal* (category A),
- 166 (29% of the total) are *moderately marginal* (category B),
- 137 (24% of the total) are *not marginal* (category C).

From a territorial point of view, the distribution appears balanced: 274 POIs are located in inland areas, while 297 are located outside them. Analyzing the demographic dimension, 163 of the urban centers identified have fewer than 5000 inhabitants, while 289 have more than 10,000 inhabitants. Finally, 303 POIs have seen a loss of young population of more than 10% over the last ten years. The results for each cycle route are shown in Table 4.

This initial analysis confirms the high potential of cycling tourism as a tool for widespread territorial regeneration. The inclusion of strategic cycling tourism infrastructure, such as CSs, can generate virtuous dynamics in marginal territories, encouraging the permanence of services, economic activities, and forms of sustainable tourism.

Table 4 Description of the Municipality and POI that belong to each cycleway

Cycleway	Crossed municipalities	POIs considered	POI categories					
			Cat. A		Cat. B		Cat. C	
			#	%	#	%	#	%
Trieste-Lignano Sabbiadoro-Venice	30	47	15	32	23	49	9	19
Sole	73	113	46	40	30	27	37	33
Tyrrhenian	97	119	47	40	35	29	37	31
Adriatic	78	72	23	32	18	25	31	43
Apulian Aqueduct	62	32	22	70	5	15	5	15
Magna Grecia	85	89	62	70	24	27	3	3
Sardinian	110	98	50	51	29	29	19	20
Total	535	571	268	47	166	29	137	24

Source Developed by the authors

4.2 Defining POIs Score

To calculate the overall score for each POI, it was necessary to define the weights to be assigned to each criterion included in the model. The values adopted are derived from the study by Pirolo and Moscarelli (2025), which provided a solid methodological basis for calibrating the parameters.

The main objective in defining the weights was to favor the selection of POIs belonging to category A, i.e., those equipped with essential services capable of more effectively supporting territorial regeneration processes.

Consequently:

- The weight w_3 was set at a very high value to exclude POIs lacking essential services from the optimal solution.
- The weight w_1 was kept low, as the preliminary selection of POIs had already taken this aspect into account.
- The weight w_7 also took on high values to increase the probability that category A POIs would be included among those chosen by the model.
- The weights w_4, w_5, and w_6 have been set to 0.3 to balance the contribution of secondary services, which can vary significantly between large centers and small villages.

Table 5 shows the details of the weights adopted for each parameter, highlighting the role of each criterion in the evaluation process.

The main objective of the weighting system adopted was to give greater importance to categories A and B than to category C, so that the overall score would give greater weight to the POIs considered most relevant to the study. Table 6, which shows the average score for each category along the different cycle routes, allows the effectiveness of the model to be verified in concrete terms. Analysis of the values shown in the table reveals that, for several cycle routes, categories A and B actually obtain higher average scores than category C, confirming the consistency of the weighting system with the methodological approach. For example, along the Trieste–Lignano Sabbiadoro–Venice cycle route, the average score for category A (5.31) and category B (5.63) is both higher than that for category C (4.60), showing how the

Table 5 Mathematical model weights according to each criteria

Criteria	Weight
k1	0.1
k2	0.9
k3	1
k4	0.3
k5	0.3
k6	0.3
k7	0.5

Source Developed by the authors

Table 6 Description of the POI average score according to the Marginality and to each cycleway

Cycleway	POIs considered	Average score		
		Cat. A (pt)	Cat. B (pt)	Cat. C (pt)
Trieste-Lignano Sabbiadoro-Venezia	47	5.31	5.63	4.6
Sole	113	6.09	4.14	5.58
Tyrrhenian	119	5.37	5.3	5.68
Adriatic	72	6.09	5.55	5.64
Apulian Aqueduct	32	4.82	5.47	–
Magna Grecia	89	5.00	4.78	5.67
Sardinian	98	5.18	4.87	3.82

Source Developed by the authors

model favors priority categories. However, there are also cycle routes where category C has higher average values than the other categories. This is not a critical issue with the model, but a reflection of the actual qualitative distribution of POIs across the territory. In some contexts, such as on the Tyrrhenian cycle route, category C (5.68) slightly exceeds A (5.37) and B (5.30), indicating that POIs of this type have a higher quality or concentration, sufficient to compensate for the lower weight assigned to this category.

Overall, the results shown in the table demonstrate that the weighting system achieves its purpose: it enhances categories A and B, while maintaining good sensitivity to the actual distribution of POIs. The model, therefore, does not impose an artificial effect but provides a balanced representation that is consistent with the territorial reality of the cycle routes analyzed.

4.3 *Specifying Technical Constraints*

After defining the various battery constraints associated with the morphological characteristics of each area, the resulting values are summarized in Table 7. The table reports the range associated with each cycle route included in the SNCT, providing an overview of the calibration adopted for the model. The Italian territory encompasses markedly different morphologies. The flattest routes are the Trieste–Lignano Sabbiadoro–Venice cycleway, planned along the northern Adriatic coast, and the Apulian Aqueduct cycleway, which develops across a uniformly flat region following the main Apulian water channel. Conversely, the Adriatic and Tyrrhenian cycleways, although also located along the coasts, present a more undulating orographic profile, which influences their total elevation gain.

From a morphological perspective, the Sole cycleway traverses the Apennines; therefore, its expected total elevation gain could be comparatively high. However, a closer analysis of its longitudinal profile reveals that it also crosses two extensive

Table 7 Average gain, chosen scenario, level of assistance, autonomy, budget, and planned CSs for each cycleway of the SNCT

Cycling route	Avg. gain (%)	Scenario	Level of assistance	Autonomy (km)	Bdg (k€)	#CS
Trieste-Lignano Sabbiadoro-Venice	0.05	Plain	ECO	122	30	15
Sole	0.29	Hill	TOUR	88	50	25
Tyrrhenian	0.31	Hill	TOUR	88	50	25
Adriatic	0.22	Hill	TOUR	88	50	25
Apulian Aqueduct	0.10	Plain	ECO	122	30	15
Magna Grecia	0.69	Mountain	TURBO	70	60	30
Sardinian	0.21	Hill	TOUR	88	50	25

Source Developed by the authors

flat areas—the Po Valley and the Florence basin—which substantially reduce the overall gain. The Sardinian cycleway exhibits a distinctly hilly profile: it spans the entire region from the coastal belt to the inland areas, which are characterized by more irregular terrain. Finally, the Magna Grecia cycleway shows the highest average elevation gain, as it follows the profile of the Calabrian Apennines.

Based on the average elevation gain, we assign each route to a specific scenario and assistance level, as described in Sect. 4.3. The Trieste–Lignano Sabbiadoro–Venice and Apulian Aqueduct routes, associated with a flat scenario, yield an autonomy of 122 km. The Sole, Adriatic, Tyrrhenian, and Sardinian cycleways fall under the hilly scenario and result in an autonomy of 88 km. Lastly, the Magna Grecia cycleway, characterized by the highest elevation gain, corresponds to the Turbo assistance level, with an autonomy of 70 km.

The results obtained from applying the mathematical model to the SNCT cycle paths take into account all the technical and territorial constraints defined in the previous phases.

In particular, the unit cost of installing each CS was set at €2000, while the total available budget was varied within a range of between €30,000 and €60,000. According to the autonomy chosen for the specific cycleway, a different number of CS is planned to be installed on the network. This choice made it possible to keep the problem computationally feasible and to analyze different realistic investment scenarios. The specific budget values adopted and the number of CSs planned for each cycle path are shown in Table 7.

Table 8 Bicycle autonomy, defined budget, number of charging stations, and POIs divided by marginal score for each cycleway of the SNCT

Cycleway	Autonomy (km)	Bgt (k€)	#CS	POI categories					
				Cat. A		Cat. B		Cat. C	
				#	%	#	%	#	%
Trieste-Lignano Sabbiadoro-Venice	122	30	14	4	27	9	60	1	13
Sole	88	50	25	10	40	6	24	9	36
Tyrrhenian	88	50	22	13	59	6	27	3	14
Adriatic	88	50	25	12	48	4	16	9	36
Apulian Aqueduct	122	30	15	7	47	8	53	0	0
Magna Grecia	70	60	30	11	36	17	57	2	7
Sardinian	88	50	25	15	60	8	32	2	8
Total	–	320	156	72	46	58	37	26	17

Source Developed by the authors

4.4 The Final Results

Analysis of the results shows that the optimal solution identified by the model does not exclusively include POIs belonging to category A, even though this represents the ideal scenario in terms of service completeness and regenerative potential.

On the contrary, the model returns a mixed selection of POIs belonging to different categories, in which, however, the sum of category A and B POIs prevails over those in category C.

This distribution highlights the model's ability to effectively balance aspects of territorial attractiveness, service availability, and economic constraints, ensuring optimal solutions even in contexts characterized by limited resources. The summary results for each cycle route are presented in Table 8, which summarizes the number and type of POIs selected, as well as the budget values used. Note that for the Trieste-Lignano Sabbiadoro-Venice Cycle Route and the Tyrrhenian Cycle Route, the total number of POIs selected is less than the available. Figure 2 shows the corresponding optimal configurations for the positioning of CS.

5 Conclusions

E-technologies can positively influence the development of slow tourism by enriching sustainable practices to experience the territory. The paper reflects on how the project of a slow tourism path can integrate services for e-mobility by planning the location of the e-bike CSs. This aspect can become a strategic issue both to make the cycleway more attractive and to encourage slow tourists to discover the places close to the

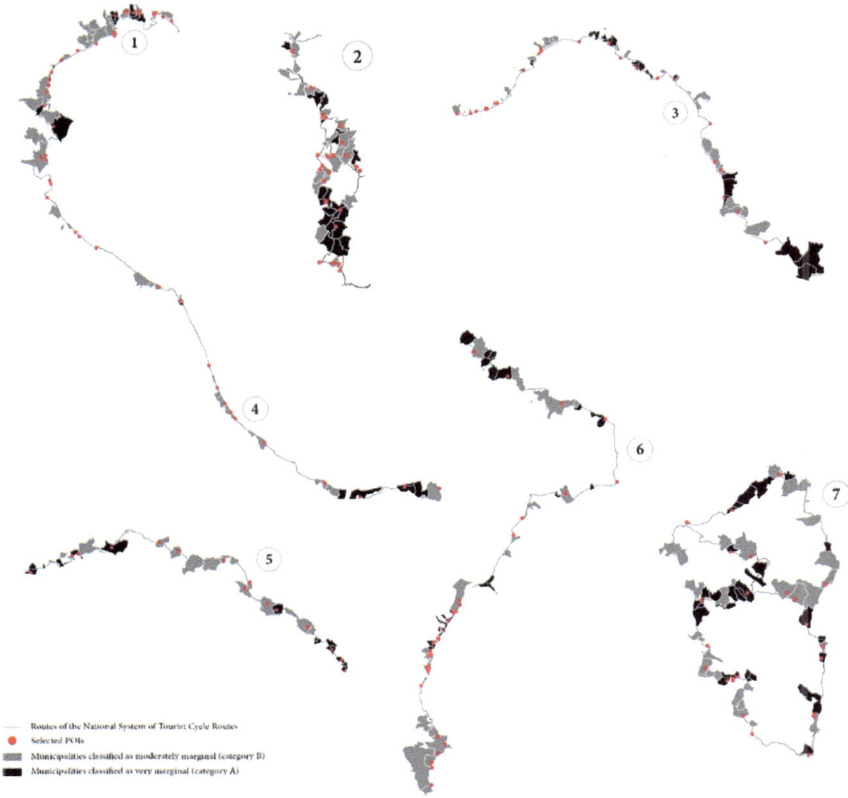

Fig. 2 Results of the of e-bike CSs location. *Source* Elaboration by authors

path, aiming to associate the technological improvements of the EVSE location with a regenerative idea of the territory. Making the location of the e-bike CSs a local development strategy was the first intention of the study. For this reason, a tool to support planning decisions, namely a mathematical optimization model in the GIS environment, has been elaborated. The idea of planning the charging infrastructure for the cycling tourist field was applied to the Italian SNCT, composed of seven itineraries. The Vento case study was already the subject of the study in Pirolo and Moscarelli (2025).

It is possible to point out that the optimization model has at least two relevant results: the integration of both quantitative and qualitative criteria in a single tool supported by a land use approach with a GIS-based spatial analysis; the replicability and simplicity of the process (e.g. by using an open data set for the GIS-based spatial analysis).

In addition to these positive aspects, it is also appropriate to consider the limits of the mathematical model, in particular related to the nature of the data. The data utilized are open-access, which is beneficial as it allows the model to operate without substantial challenges or costs related to data acquisition. However, this advantage is counterbalanced by the absence of significant and detailed information, particularly concerning the characterization of services in smaller towns. The database used lacks the depth required to accurately depict the situation in these areas. Leveraging a digital mapping service API, such as Google Maps, could provide richer data on commercial activities and essential community services, including post offices, clinics, and schools. Another limitation is the weighting score. In fact, in the analyzed context, there is a wide variety of towns, and a more in-depth analysis could lead to obtaining better results that reflect reality. Concerning the mathematical model, we note that it still focuses just on individual travel scenarios, avoiding the added complexity of the group travel analysis. It would imply actors like synchronization, where individuals within a group must align their schedules to visit shared points of interest, and diversity in preferences, which requires balancing common and individual goals. Other challenges could include managing varying budgets, fitness levels, and time constraints, which necessitate a more flexible and dynamic approach to itinerary planning. Future developments could thus incorporate algorithms to handle the partitioning of groups into subgroups with overlapping or distinct itineraries, optimizing for shared experiences while preserving personalization. Such extensions would harmonize individual preferences with collective goals, adding layers of complexity to the optimization process but significantly broadening the model's applicability and relevance.

Despite these limitations, the proposal remains interesting, especially for its general intention of integrating slow tourism with EVSE services to foster the sustainable development of marginal areas. E-mobility not only triggers and enhances slow tourism but can also contribute to generating positive economic impacts if the location of the e-bike CS is defined not only by technical and budget constraints but also by other spatial criteria. Such a perspective is quite original and can be regarded as a valid result of the research. Moreover, the premises at the basis of the model promote local development of less attractive places, which suits with public interest.

This also opens to a first reflection on some managerial implications that are not the main object of the paper, but that can become starting points for the future development of the research. As a matter of fact, the model has been designed as a practical tool for the public institutions and local authorities involved in slow tourism projects. It would be necessary to better define the entire process to integrate e-services with slow infrastructures: from the planning of e-bike CSs to their installation and maintenance over time. From this perspective, the model could be seen as a first planning tool, useful to orient the public actor. The other phases should be more shared with private stakeholders (e.g., e-bikes CSs providers and producers or local actors involved in accommodation activities, which can be responsible for the maintenance of the EVSE services), even for what concerns the funding aspects.

Finally, the study is a preliminary step in the analysis of this new and complex development for the slow tourism field and can be improved both by trying to overcome the technical limitations of the model and by considering a wider view of all the different phases.

Acknowledgements The authors have shared the concept of the article. P. L. B. supervised the development of the model. L.P. wrote Sect. 3 and 4. R.M. wrote Sects. 1 and 2. R.M. wrote Sects. 1, 2. All together, the authors wrote Sect. 5.

Author Contributions The paper is a result of the study carried out within the MOST—Sustainable Mobility National Research Center and received funding from the European Union Next-Generation EU (PIANO NAZIONALE DI RIPRESA E RESILIENZA (PNRR)—MISSIONE 4 COMPONENTE 2, INVESTIMENTO 1.4—D.D. 1033 17/06/2022, CN00000023). This chapter reflects only the authors' views and opinions; neither the European Union nor the European Commission can be considered responsible for them.

References

Behrendt F (2018) Why cycling matters for electric mobility: towards diverse, active and sustainable e-mobilities. Mobilities 13:64–80

Belotti P, Errico F, Malucelli F, Massetti AT (2024). Optimization of E-bike networks. Transp Plan Technol 1–22

Bosch eBike Systems (2025) Assistente Autonomia eBike. https://h7.cl/1ffDW. Accessed 27 Nov 2025

Burani E, Cabri G, Leoncini M (2022) An algorithm to predict e-bike power consumption based on planned routes. Electronics 11

Černá A, Černý J, Malucelli F, Nonato M, Polena L, Giovannini A (2014). Designing optimal routes for cycle-tourists. https://doi.org/10.1016/j.trpro.2014.10.064

Dong G, Ma J, Wei R, Haycox J (2019) Electric vehicle charging point placement optimisation by exploiting spatial statistics and maximal coverage location models. Transp Res Part d: Transp Environ 67:77–88

Gazzola P, Pavione E, Grechi D, Ossola P (2018) Cycle tourism as a driver for the sustainable development of little-known or remote territories: the experience of the Apennine regions of Northern Italy. Sustainability 10:1863. https://doi.org/10.3390/su10061863

Han H, Lho LH, Al-Ansi A, Yu J (2020) Cycling tourism: a perspective article. Tourism Rev 75:162–164

Lagerstedt E, Svensson H (2022). Do tourists dream of electric bikes?: Electric bikes as a mean to improve sustainability of tourism in rural Sweden. In: Proceedings of the 25th international academic Mindtrek conference, pp 167–178

Massetti AT (2020) The e-bike charger location problem. https://hdl.handle.net/10589/188353 Accessed 12 Dec 2025

Micari S, Polimeni A, Napoli G, Andaloro L, Antonucci V (2017) Electric vehicle charging infrastructure planning in a road network. Renew Sustain Energy Rev 80:98–108. https://doi.org/10.1016/j.rser.2017.05.022

Petroman I M, Văduva L, Marin D (2021). Benefits of slow tourism for local economies. Quaestus 280–291

Pileri P, Moscarelli R (2021) Cycling & walking for regional development. Springer, Switzerland

Pirolo L, Belotti P, Malucelli F, Moscarelli R, Pileri P (2024) Optimal charging station location in a linear cycle path with deviations. In: International symposium on combinatorial optimization. Springer, pp 361–375

Pirolo L, Moscarelli R (2025). Integrating cycling tourism, e-mobility and local development: a spatial model to plan slow tourist paths for e-bikes. Regional Sci Policy Practice 17(12)

Pröbstl-Haider U, Lund-Durlacher D, Antonschmidt H, Hödl C (2018) Mountain bike tourism in Austria and the Alpine region–towards a sustainable model for multi-stakeholder product development. J Sustain Tour 26:567–582

Senese A, Pelfini M, Maragno D, Bollati IM, Fugazza D, Vaghi L (2023). The role of e-bike in discovering geodiversity and geoheritage. Sustainability 15

UVAL (2014) Public investment evaluation unit. https://www.agenziacoesione.gov.it/wpcontent/uploads/2020/07/MUVAL_31_Aree_interne_ENG.pdf. (accessed 20 December 2024)

Vansteenwegen P, Van Oudheusden D (2007) The mobile tourist guide: an OR opportunity. Or Insight 20:21–27

Verbeeck C, Vansteenwegen P, Aghezzaf E-H (2014) An extension of the arc orienteering problem and its application to cycle trip planning. Transp Res Part e: Logist Transp Rev 68:64–78. https://doi.org/10.1016/j.tre.2014.05.006

Zhu ZH, Gao ZY, Zheng JF, Du HM (2016) Charging station location problem of plug-in electric vehicles. J Transp Geogr 52:11–22

Optimal Location of E-Bike Chargers on Touristic Cycleway Networks

Luca Pirolo, Pietro Belotti, and Federico Malucelli

Abstract E-bikes allow for longer and easier rides even for untrained cyclists, but the riders can take full advantage of their bikes on cycleway networks that are equipped with Charging Stations (CSs) for extended autonomy. Therefore, CSs must be deployed on a cycleway network so that a bicycle rider can find a CS before running out of charge, though budget constraints limit the number of CS that can be installed. In cycleway networks structured as simple paths, we prove that the problem of deciding where to install CSs in a uniform way and complying with budget constraints can be mathematically formulated and solved very efficiently with an elegant algorithm supported by a graph representation of the problem. The approach will be extended to linear cycleways having alternative branches. The more general case of a cycleway network is studied. This case requires a more complex discrete optimization model, and alternative approaches will be discussed. The models will be applied to a set of cycleway networks with different shapes.

1 Introduction

Cycleway networks are now more accessible to the general public thanks to assisted-pedaling bicycles (e-bikes), which carry a battery-powered electric motor that increases the distance autonomy of the rider and facilitates ascents, thus allowing for longer, more enjoyable rides. Riders with various training levels can therefore explore vast cycleway networks, even in mountain areas, for hours without significant fatigue and with limited environmental impact.

L. Pirolo (✉) · P. Belotti · F. Malucelli
Department of Electronics, Information, and Bioengineering—Politecnico di Milano, Milan, Italy
e-mail: luca.pirolo@polimi.it

P. Belotti
e-mail: pietro.belotti@polimi.it

F. Malucelli
e-mail: federico.malucelli@polimi.it

© The Author(s), under exclusive license to Springer Nature Switzerland AG 2026
R. Moscarelli et al. (eds.), *The Future of Slow Mobility*, Research for Development,
https://doi.org/10.1007/978-3-032-23660-9_14

E-bikes have batteries of various sizes, and the battery's autonomy is sometimes insufficient to guarantee long rides of more than a few tens of kilometers, thus limiting the exploration of vast areas. However, their batteries can be recharged by plugging them into charging stations (CSs) along the routes, extending the range of a bicycle ride and hence the attractivity of e-bike tourism. Charging stations differ from those for electric vehicles in that they require much less power and can be installed even in remote areas by adding a photovoltaic panel and accumulators to provide energy. The charging time, typically one to two hours, requires the cyclist to pause the ride, but also allows for exploring the local environment of the CS and its attractions.

Therefore, the decision maker aiming to make a cycleway network attractive should install CSs in the most adequate locations, ensuring accessibility for all e-bike riders while respecting budget constraints and maintaining a uniform maximum distance between consecutive CSs. This is an optimization problem, and its mathematical formulation and algorithmic solution have been the subject of several research efforts in the broader field of cycleway network design.

While in the scientific literature and in the digital tools available on the web or on portable devices we can find a good variety of contributions regarding route generation using optimization tools, the contributions on the decision making choices to design and upgrade cycleway networks are not so frequent, especially when touristic cycleways are considered.

An early reference in this area is (Černá et al. 2014), which addresses the problem of designing optimal routes for cycle-tourists. Their model integrates multiple criteria as distance, safety, and attractiveness, into a route optimization framework, demonstrating how mathematical optimization can produce itineraries that balance efficiency and recreational value. This study represents a pivotal step toward formalizing bicycle route planning as an optimization problem and has inspired subsequent works addressing larger-scale network planning. Giovannini et al. (2017) further extends these concepts by introducing a design framework for cycle-tourist networks at the regional level, where the objective is to identify optimal connections among existing routes and points of interest.

Later research extended these ideas from individual routes to the design of entire cycleway networks. Mauttone et al. (2017) proposed a mixed-integer optimization model for selecting optimal links in an urban bicycle network, considering both construction costs and user equilibrium flows. Such approaches aim to identify the subset of links that maximizes network performance or accessibility within budget constraints, often accounting for the way cyclists choose routes once the network is built. Similar formulations have been used to plan network expansion in developing cities or regions with evolving cycling demand.

A key advancement has been the integration of behavioral choice models into optimization frameworks. Liu et al. (2019) introduced a path-size logit model within a network design formulation to capture overlapping and competing routes, providing more realistic demand assignment. Other studies have expanded the objective functions to include multiple criteria as safety, connectivity, comfort, and cost, resulting in multi-objective optimization models solved via heuristic or metaheuristic algorithms.

Recent contributions have focused on the economic efficiency and social welfare implications of cycling infrastructure. Caicedo et al. (2023) analyzed cost-effective design parameters under budget constraints, while Paulsen and Rich (2024) proposed a welfare-optimal network expansion framework that incorporates induced demand and accessibility benefits. These approaches emphasize the role of optimization not only as a technical design tool but also as a decision-support framework for sustainable transport policy.

Other recent contributions adopt data-driven approaches: Steinacker et al. (2022) propose a demand-driven optimization framework using empirical cycling data to improve urban "bikeability" while Lonardi et al. (2025) apply optimal transport theory to achieve cohesive and well-connected cycleway networks in multilayer urban environments. These approaches represent a shift toward data-informed, adaptive optimization models capable of integrating diverse performance metrics as connectivity, safety, accessibility, and energy consumption.

A wide range of solution approaches has been explored, from exact mixed-integer linear programming for small-scale networks to heuristic and metaheuristic techniques for larger instances. Decomposition algorithms, greedy link-addition strategies, and hybrid optimization pipelines have been successfully applied to handle the combinatorial complexity of real-world cycleway networks. Data-driven methods and GIS-based preprocessing steps are increasingly used to integrate spatial, topographic, and environmental information into the optimization process (Natera Orozco et al. 2020).

Emerging research focuses on robust and adaptive network design, accounting for uncertainties in demand, weather, and future technological developments. Several studies highlight the need to explicitly incorporate topography and energy consumption for e-bikes, especially in hilly or mountainous regions, and to model charging infrastructure placement jointly with network optimization. This integration is particularly relevant for e-bike tourism, where route feasibility depends on the spatial distribution of CSs and on the riders' autonomy limits (Burani et al. 2022).

Open challenges include the scalability of exact optimization methods, the empirical calibration of user behavior parameters, and the integration of multimodal interactions (e.g., cycling combined with public transport). Addressing these aspects will enhance the realism and applicability of optimization-based models for cycleway network planning, particularly in the context of e-bike tourism and sustainable regional mobility.

In this chapter we consider the problem of upgrading a cycleway network by installing e-bike chargers in strategic positions, so as to improve the accessibility and to attract tourists in remote areas. We build on the preliminary models proposed in Belotti et al. (2024), specializing them to the case of linear cycleways for which we derive efficient solution methods. The method developed for linear cycle network is then extended to general networks. Considering the touristic nature of the network, we assume that the capacity of the links and of the chargers is not an issue, and that a specific demand of transportation does not exist. The idea is that we do not introduce the chargers to follow the demand, but rather the opposite. The presence of the chargers will create the demand, and the their position must be decided so as to cover uniformly the network, minimizing the maximum distance between two consecutive chargers in any possible itinerary. The public and private stakeholders that are often involved in the development of CS infrastructure interact in such a way that initial public investment in CSs spurs more private investments and attracts demand.

The chapter is organized as follows. We first introduce the problem and the basic notation (Sect. 2), then describe an efficient solution method for the linear cycleway case (Sect. 3). The method is extended to the linear case with two alternative branches (Sect. 4) and finally generalized to the case of cycle-networks (Sect. 5). Some results obtained by applying the proposed method to some Italian and European cycleways and network are illustrated in Sect. 6.

2 Optimally Locating Charging Stations on Cycle Paths

In this section we present the general problem and we introduce the notation that will be used in the chapter. Consider a network of bike trails represented as a directed graph $G = (N, A)$, where the set of nodes N is the set of pertinent sites and the set of edges A gives the possible direct connections between pairs of sites. A distance d_{ij} is associated with every edge $\{i, j\}$ of A. Let us suppose that S is the subset of nodes where a CS can be installed, and the installation cost at node $i \in S$ is denoted by c_i. The maximum budget for the CSs is F. In addition, we suppose that a set P of possible riders' itineraries is given. Each itinerary $p \in P$ is specified by a sequence of nodes $(i_1, i_2, \ldots, i_{m_p})$ that defines the sequence of sites such that consecutive sites have a direct connection.

In Fig. 1 we illustrate a network with nine nodes and 3 itineraries, together with a possible solution for the CS location.

Given a set S of nodes candidates for installing a CS and one itinerary p, we define a *stretch* as a portion of itinerary between two consecutive candidate sites in S. Formally, a stretch is identified by a pair $(i_{k'}, i_{k''})$ of nodes from p such that $1 \leq k' < k'' \leq m_p, i_{k'} \in S, i_{k''} \in S$, and $i_k \in N \setminus S, \forall k : k' < k < k''$. For simplicity, we assume that $i_1, i_m \in S$. The *stretch length* of (i'_k, i''_k) is defined as the total distance on p between $i_{k'}$ and $i_{k''}$, i.e., $\sum_{k=k'}^{k''-1} d_{i_k, i_{k+1}}$. Finally, the *longest stretch* on path p is the stretch whose length is maximum.

Locating CSs on the nodes of the network while minimizing the maximum stretch length and ensuring that the total CS installation cost is at most F is an optimization problem that we call *Capacitated MinMax Stretch Problem* (CMMSP) as it is introduced in Pirolo et al. (2024). The question posed by this problem is the following: in what nodes of the paths $p \in P$ should a CS be installed so that (i) installation cost is within budget and (ii) the maximum distance that a cyclist may have to travel, without running out of battery, while entering and exiting the graph at any point (intermediate or not) of the path p, is minimum?

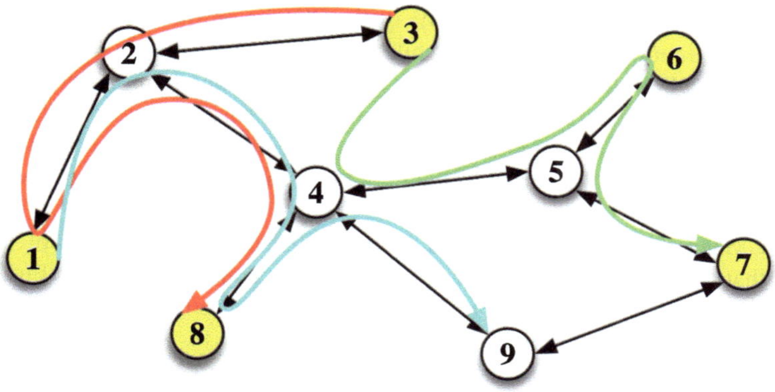

Fig. 1 Example of a cycleway network with subset $S = \{1, 3, 6, 7, 8\}$ for the candidate CSs and itineraries $p_1 = \{3, 2, 1, 2, 4, 8\}$, $p_2 = \{1, 2, 4, 8, 4, 9\}$, $p_3 = \{3, 4, 5, 6, 5, 7\}$. The yellow nodes represent the decisions about the sites to be equipped with a charging station

In Sect. 5 we will provide a Mixed Integer Linear Optimization model for the general CMMSP. But first we start analyzing the special case of a linear cycleway and its variants. These cases can exploit the special structure of the problem that allows for an efficient algorithmic approach yielding the optimal solution in very short time. The general algorithmic approach will be extended to the general case.

3 The Case of a Linear Network and One Itinerary

Before considering the most general case, we studied the case of a linear cycleway. This type of network is quite common, thinking about the cycleways following the big rivers. Moreover, we could have easy access to the data of the Venezia-Torino (VenTo) cycleway (Pileri 2020) to test our model and algorithm.

In order to induce tourists to discover the territory, it is not recommended to install CS directly on the main path of the cycleway. Instead, ideally, they should be installed in some sites in the proximity of the main course, called Points of Interest (POI). The motivation is that the presence of a CS induces riders to deviate from the main course and discover the territory (Pirolo and Moscarelli 2025, 2026). Moreover, especially in the case of cycleways following the river banks, the possibility of carrying out leisure activities on river banks during the charging operation is rare, while deviating from the main course to reach villages or facilities offers a variety of alternative activities that justify the stop.

The graph representing the case of a linear cycle path is thus a so-called *caterpillar* (see Fig. 2), i.e., a graph that resembles the structure of a cycle path where each node is connected to one or more nearby points of interest (POIs), and the POIs are connected among them through the linear cycle path.

In summary, we assume that (i) none of the POIs is on the main cycle path and that (ii) to facilitate leisure activities while recharging the e-bike, a CS should only be installed at a POI, therefore only nodes that do not belong to the caterpillar spine can host a CS. Finally, in order to include the extreme case where a cyclist visits all POIs between her entrance and exit nodes, we consider a path p that begins at the first node of the caterpillar graph's spine and ends at the last one, and it visits all intermediate ones.

The CMMSP on a cycle path that is connected to a set of points of interest implies that $G = (N, E)$ is structured as a caterpillar graph, as shown in Fig. 2. Let $N = L \cup S$ where $L = \{1, 2, \ldots, n\}$ is a set of nodes that correspond to the locations along the cycle path from which we can deviate, and $S = \{n+1, n+2, \ldots, 2n\}$ is a set of nodes corresponding to the POIs that may host a CS. The set of edges E is the union of two subsets: $E' = \{\{i, i+1\}, i = 1, 2, \ldots, n-1\}$, i.e., the edges forming the cycle path proper; and $E'' = \{\{i, n+i\}, i = 1, 2, \ldots, n\}$, which form the connection between each point i on the cycle path and a nearby POI.

The distances between consecutive nodes in L (i.e., $d_{i,i+1}, i = 1, \ldots, n-1$) and the lengths of the deviations ($d_{i,n+i}, i = 1, \ldots, n$) are given. Due to the practical considerations outlined before, we also assume that the path p goes from the first to the last extreme of the cycleway and touches all POIs in between.

An optimization model can be developed by considering a secondary caterpillar graph with two extra nodes called α and ω and with edges $\{\alpha, 1\}$ and $\{n, \omega\}$, both having null distance, as shown in Fig. 3. Path $p = (\alpha, 1, n+1, 1, 2, n+2, 2, 3, \ldots, n-1, n, 2n, n, \omega)$ is indicated with a dashed curve through all nodes from α to ω, traversing each node in $\{1, 2, \ldots, n\}$ twice.

We introduce an augmented graph $G^* = (N^*, A^*)$ as the directed acyclic graph defined with $N^* = \{\alpha, 1, 2, \ldots, n, \omega\}$ and $A^* = \{(i, j) : i, j \in N^*, i < j\}$, where, by a little abuse of notation, we assume $\alpha < i < \omega$ for all $i = 1, 2, \ldots, n$. Define

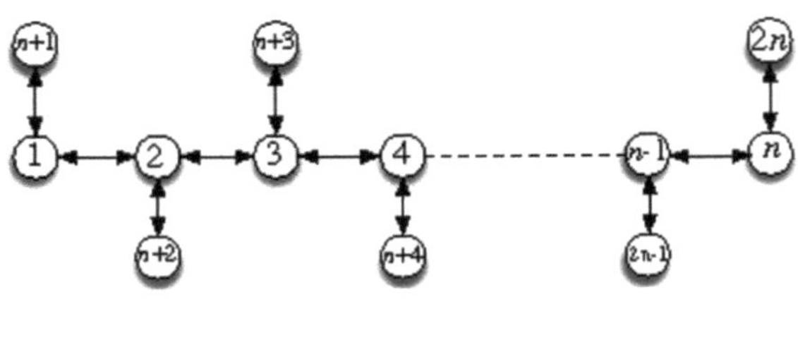

Fig. 2 Example of a caterpillar graph deriving from a linear cycleway: $1, 2, \ldots, n$ are the nodes forming the cycle path, whereas $n+1, n+2, \ldots, 2n$ are the nodes corresponding to the points of interest, where CSs can be installed

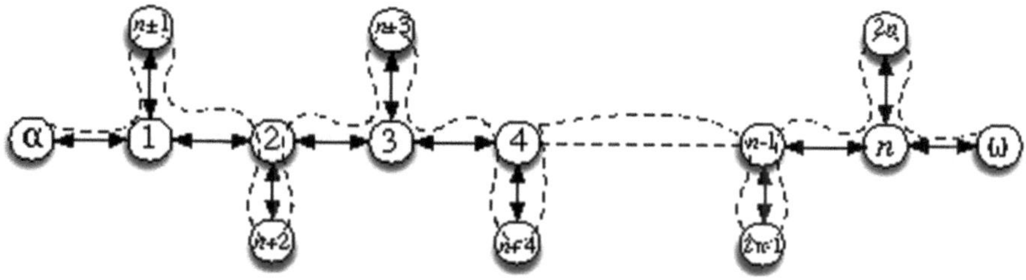

Fig. 3 Secondary caterpillar graph with fictitious source and destination nodes α and ω. The dashed curve is the cyclist trajectory p through all POIs

now the traveling length l_{ij} on all $(i, j) \in A^*$ as the total distance of an $i - j$ path that traverses once every POI node in between:

$$l_{ij} = \begin{cases} \sum_{h=1}^{j-1} d_{h,h+1} + 2\sum_{h=1}^{j-1} d_{h,n+h} + d_{j,n+j} & \text{if } i = \alpha \\ d_{i,n+i} + \sum_{h=i}^{n-1} d_{h,h+1} + 2\sum_{h=i+1}^{n} d_{h,n+h} & \text{if } j = \omega \\ d_{i,n+i} + \sum_{h=i}^{j-1} d_{h,h+1} 2\sum_{h=i+1}^{j-1} d_{h,n+h} + d_{j,n+j} & \text{otherwise} \end{cases} \tag{1}$$

Every arc $(i, j) \in A^*$ represents a stretch on G^* with length l_{ij}. Arc (i, j) is associated a label pair (l_{ij}, f_{ij}), where l_{ij} is defined by (1), while the arc cost f_{ij} is the cost c_{n+j} of installing a CS at the head node j of the arc, with $f_{i,\omega} = 0, \forall i \in N^*$. Note that, by construction, G^* is an acyclic graph, since for all arcs (i, j) we have that $i < j$.

An example for a cycle path with $n = 4$ and for edge distances $d_{i,i+1} = 5, \forall i = 1, 2, \ldots, n - 1$ and $d_{i,n+i} = 2, \forall i = 1, 2, \ldots, n$ is provided in Fig. 4. Note that in general, for a cycle path with n exit points and consequently n points of interest, the augmented graph contains $O(n^2)$ arcs.

We reduce the CMMSP to the problem of finding an $\alpha - \omega$ path p^* whose total cost, described with arc costs f_{ij}, is below the given budget F and such that $\max_{(i,j)\in p^*} l_{ij}$ is minimum. An optimal solution to CMMSP is then given by all nodes $n + j \in N$ corresponding to arcs $(i, j) \in p^*$. Let us consider two sample solutions on the graph in Fig. 4, where for simplicity we use c_i in place of c_{n+i}. The first one is defined by the path $(\alpha, 2, \omega)$, equivalent to installing a single CS at node 2; its two stretches $\alpha - 2$ and $2 - \omega$ have length $2 + 2 + 5 + 2 = 11$ and $2 + 5 + 2 + 2 + 5 + 2 + 2 = 20$, respectively, hence the maximum stretch length is 20 while its CS installation cost is given by $c_2 + 0 = c_2$. Consider a second solution given by path $(\alpha, 1, 2, 4, \omega)$, equivalent to installing a CS at nodes 1, 2, and 4. The stretch lengths are therefore 2, 2 + 5 + 2 = 9, 2 + 5 + 2 + 2 + 5 + 2 = 18, and 2, their maximum being 18, while the total installation cost is c1 +c2 +c4. Clearly the second solution has better objective function (maximum stretch length), but it has a larger installation cost which may be infeasible if $c_1 + c_2 + c_4 > F$.

In order to formulate the problem mathematically, we introduce the binary variables x_{ij} for each $(i, j) \in A^*$ that are set to 1 if the stretch $i - j$ is part of the solution.

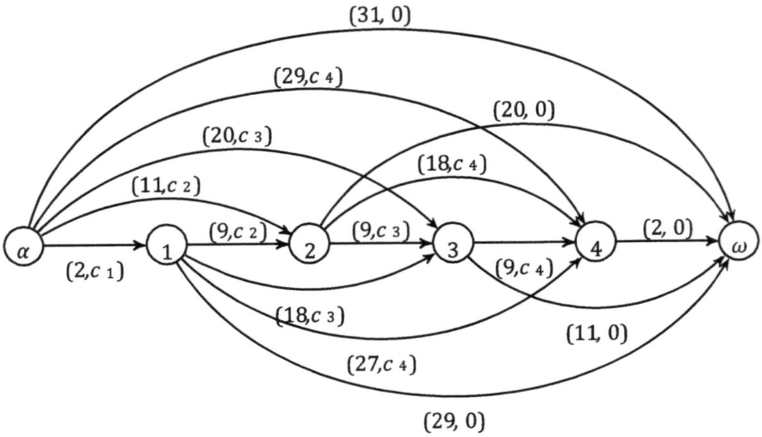

Fig. 4 Augmented graph for a sample caterpillar graph with $n = 4$ and edge distances $d_{i,n+i} = 2$ for $i = 1, 2, 3, 4$, $d_{1,2} = d_{2,3} = d_{3,4} = 5$, and $d_{\alpha,1} = d_{4,\omega} = 0$. Each arc has label (l_{ij}, f_{ij}) indicating the stretch length and the installation cost

Moreover we need a bottleneck variable δ to account for the length of the longest stretch. The problem can be formulated as a resource constrained bottleneck path problem:

$$\text{CMMSP}: \quad \min \delta \tag{2}$$

$$l_{ij}x_{ij} \leq \delta \qquad \forall (i, j) \in A^* \tag{3}$$

$$\sum_{(i,j)\in A^*} f_{ij}x_{ij} \leq F \tag{4}$$

$$\sum_{(j,i)\in BS(i)} x_{ji} - \sum_{(i,j)\in FS(i)} x_{ij} = b_i \ \forall i \in N^* \tag{5}$$

$$x_{ij} \in \{0, 1\} \qquad \forall (i, j) \in A^* \tag{6}$$

where $b_i = -1$ if $i = \alpha$, $d_i = 1$ if $i = \omega$, and $FS(i)$ and $BS(i)$ denote the outgoing and incoming arcs in i, respectively. Constraints (3) define the value of the bottleneck variable, in conjunction with the objective function. Constraint (4) limits the total cost of the solution and constraints (5) define the paths from α to ω.

Rather than solving CMMSP using Mixed Integer Linear Optimization methods, a simple observation allows for solving the problem in polynomial time. Consider a fixed value of the maximum allowed stretch length δ, the problem becomes equivalent to determining, via a shortest path algorithm using labels f_{ij}, whether an $\alpha - \omega$ path exists whose total cost is at most F. A binary search then suffices to find the optimal value of δ. The binary search method executes at every iteration a shortest path algorithm for acyclic graphs, whose worst case complexity is linear in the number of arcs, and determines whether the budget constraint is violated (in which case the range of the binary search is decreased) or whether it is satisfied (and then the range is increased) until the optimal value of the stretch length is found.

The pseudocode of the binary search algorithm is described below.

Algorithm 1: Binary Search for CMMSP

```
Input: G = (N, A), budget F, CS costs c
Output: Optimal value of δ
Create augmented graph G* = (N*, A*) with labels l_ij, f_ij;
[δ_min, δ_max] ← [0, l_αω];
while δ_min < δ_max do
    δ ← 1/2 (δ_min + δ_max);
    A** ← {(i, j) ∈ A* : l_ij ≤ δ};
    Find the length l of the α–ω shortest path on G** = (N*, A**) with weights
    f_ij;
    if l > F then
        δ_min ← δ;
    else
        δ_max ← δ;
return δ;
```

3.1 Application to the VenTo Cycle Path

The VenTo (from VENezia and TOrino) is a cycle path about 700 kilometers long that runs along the Po River, between Turin and Venice via Milan: it crosses four regions (Piedmont, Lombardy, Emilia Romagna, and Veneto), 13 districts/provinces and 118 municipalities. In 2016, it was included in the first Italian strategy to realize ten touristic cycle routes, the so-called "Ciclovie Turistiche Nazionali" (Ministero 2018). Figure 5 shows the complete map of the cycle path. Note that the structure does not follow that of a single path but (1) it contains a long connecting trail from Pavia to Milan; and, most importantly, (2) it splits into two parallel trails between Piacenza and Cremona.

We have applied the algorithm described in the previous section to a portion of VenTo ranging from Piacenza to San Benedetto Po (province of Mantua), with a total length of about 210 km. Figure 6 shows in detail this portion, together with the selected POIs. In order to apply our algorithm, only the southern of the parallel trails between Piacenza and Cremona has been considered, which makes the portion under analysis a path. We will generalize the approach considering the two alternative branches in Sect. 4.

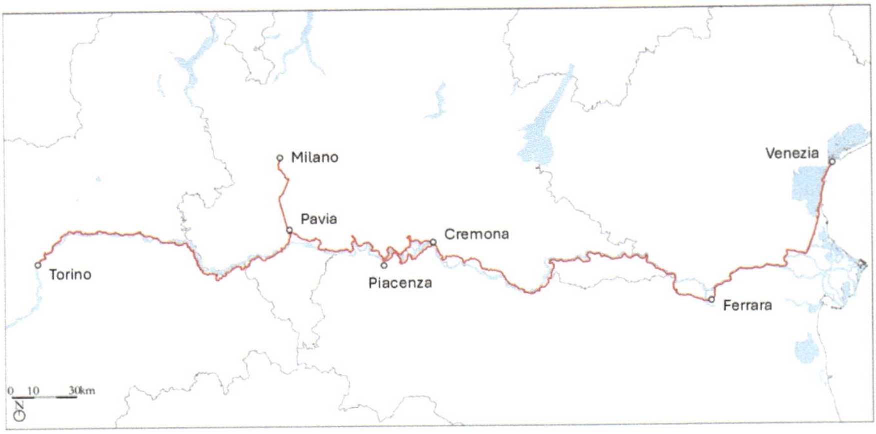

Fig. 5 The overall VenTo cycle path, from Turin to Venice

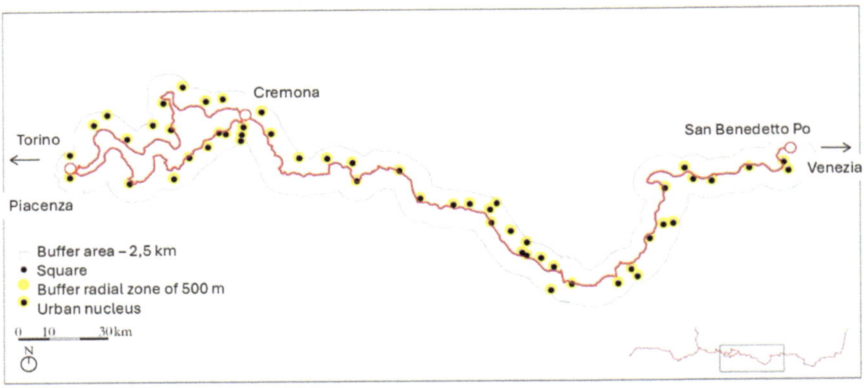

Fig. 6 Portion of the cycle path under study

The caterpillar graph has been generated by automatically extracting data with suitable GIS tools. The resulting graph contains $n = 44$ nodes on the cycle path and about as many nearby POIs. Note that some of the latter are connected to the cycle path through the same exit point: the instance can be adapted by creating multiple distinct exit points, which have a distance of 0 from one another, so that each exit point is connected independently to the corresponding POI.

The experimental tests were conducted on a Lenovo ThinkPad with a QGIS installation. All algorithms used for this work were either coded in Python 3.11.5 (binary search) or used Python modules such as NetworkX (Hagberg et al. 2008) for the graph algorithms. While CSs can be of several types, for simplicity here we have decided to assign an equal price $c_i = 2000$ for all POIs. This translates into a simplification of the budget constraint into a cardinality constraint,

Table 1 Run times, optimal value of δ and location of the CSs when the budget F varies

Budget	Time (s)	δ_{opt} (km)	Optimal set of CSs
2000	0.31	133.2	{22}
4000	0.26	89.5	{13, 29}
6000	0.22	68.4	{10, 20, 33}
8000	0.20	54.8	{7, 16, 27, 36}
16,000	0.14	33.3	{2, 10, 13, 18, 24, 30, 35, 38}
24,000	0.14	24.1	{1, 3, 9, 12, 15, 18, 23, 27, 30, 34, 37, 38}
32,000	0.12	19.0	{1, 3, 7, 11, 12, 14, 17, 20, 23, 27, 29, 31, 34, 37, 38, 40}
64,000	0.12	18.6	{1, 3, 7, 11, 12, 13, 16, 18, 22, 25, 28, 30, 32, 35, 37, 38, 40}

and therefore the shortest-path step in our algorithm consists, for these tests, in finding an $\alpha - \omega$ path with the smallest number of arcs. We now assess the performance of the algorithm on the VenTo instance for budget F ranging in $B = \{2000, 4000, 6000, 8000, 16{,}000, 24{,}000, 32{,}000, 64{,}000\}$. For each $F \in B$ we have run the binary search algorithm described above, and we report in Table 1 the run time for the binary search, the optimal value of δ and the corresponding optimal set of CSs. For $F = 8000$, for instance, the optimal solution $\{7, 16, 37, 36\}$ consists of the POIs Castelvetro Piacentino, Motta Baluffi, Viadana, and Suzzara.

4 Linear Cycle Path with Alternative Branches

A slight extension to the simple path case is one in which the graph is composed of a path and one cycle, the two having one node in common. This is the example depicted in Fig. 6, where there are two alternative disjoint paths to go from Piacenza to Cremona, the latter being the first node of the main path to San Benedetto Po and forward. We suppose that a cyclist starting from Piacenza either selects the northern side or the southern side. Then the cyclist, after reaching Cremona, continues along the unique path that goes towards Venice.

The binary search algorithm proposed in the previous section cannot be applied to this case, as the augmented graph can only be defined for linear paths.

One could solve two problems: one that considers the northern alternative with the rest of the cycleway, and one that considers the southern alternative with the rest of the cycleway. However, the two solutions may not coincide in the common parts. This implies that the two problems must be solved in conjunction and a more general optimization approach must be proposed. The mathematical model of the CMMSP must be duplicated and then the equality of the decision variable in the common portion must be imposed. Let us denote by $G_1 = (N_1^*, A_1^*)$ and $G_2 = (N_2^*, A_2^*)$ the complete augmented graphs related with the first and the second variant. The two graphs will have a common portion defined by $N_1^* \cap N_1^*$ and $A_1^* \cap A_2^*$. In the mathematical for-

Table 2 Run times, optimal value of δ and location of the CSs for different values of the budget F. In parentheses we have indicated the CS in the alternative branches separated by /

Budget	Time (s)	δ_{opt} (km)	Optimal set of CSs
6000	1.03	68.4	$\{10, 20, 33\}$
8000	1.22	65.8	$\{(8/47), 15, 28, 39\}$
12,000	1.03	45.0	$\{(5/42), 10, 15, 23, 31, 39\}$
16,000	1.12	34.2	$\{(4, 8/43)11, 15, 21, 28, 33, 39\}$
20,000	1.25	29.8	$\{(4, 7/43), 10, 13, 17, 21, 25, 30, 34, 3\}$
24,000	1.41	26.9	$\{(3, 6, 9/42, 45), 11, 13, 17, 22, 27, 31, 35\}$
32,000	1.12	21.5	$\{(3, 5, 8, 9/42, 45), 11, 12, 15, 18, 22, 26, 29, 32, 35\}$

mulation we introduce two sets of variables $x_{ij}, \forall(i, j) \in A_1^*$ and $y_{ij}, \forall(i, j) \in A_2^*$ that define the paths in the two graphs [constraints (13)]. The two path subproblems are then independent, except for the common portion of the cycle path, for which we impose the equality. The resulting mathematical formulation is as follows.

$$
\begin{aligned}
CMSSP': \quad & \min \delta && && (7)\\
& l_{ij}x_{ij} \leq \delta && \forall(i, j) \in A_1^* && (8)\\
& l_{ij}x_{ij} \leq \delta && \forall(i, j) \in A_2^* && (9)\\
& \textstyle\sum_{(i,j)\in A_1^*} f_{ij}x_{ij} + \sum_{(i,j)\in A_2^*} f_{ij}y_{ij} \leq F && && (10)\\
& \textstyle\sum_{(j,i)\in BS(i)} x_{ji} - \sum_{(i,j)\in FS(i)} x_{ij} = b_i && \forall i \in N_1^* && (11)\\
& \textstyle\sum_{(j,i)\in BS(i)} y_{ji} - \sum_{(i,j)\in FS(i)} y_{ij} = b_i && \forall i \in N_2^* && (12)\\
& x_{ij} = y_{ij} && \forall(i, j) \in A_1^* \cap A_2^* && (13)\\
& x_{ij} \in \{0, 1\} && \forall(i, j) \in A_1^* && (14)\\
& y_{ij} \in \{0, 1\} && \forall(i, j) \in A_2^* && (15)
\end{aligned}
$$

Unfortunately, the additional constraints spoil the original structure of the problem, which cannot be solved with the same method. A standard MILO solver can be used instead. Clearly the resulting solution method is less efficient than our binary search approach, but we can still find a solution in a reasonable computation time. Table 2 summarizes the results obtained in the VenTo case study.

5 Formulation of the General Problem Considering a Predefined Set of Itineraries

The general CMMSP, defined in Sect. 2, considers general graphs without any assumption on their structure. However, we assume that a set of input itineraries is given as a set of paths P defined on the graph representing the cycleway network.

These paths identify the possible cyclist trajectories, and they should span all the network, so as to guarantee a suitable service coverage. The generalization of the model described for the simple path case is a significant extension that cannot exploit any structure of the graph underneath. Thus, the path search approach described in Sect. 3 cannot be applied. The optimization model proposed for the general case considers each input path as separate in a multi-commodity setting. Among the extensions to be considered is also the asymmetric value of the length measure for a network edge when traversed in either direction: on planar cycleway networks the stretch length does not change when the traversal direction changes, hence the distance d_{ij} is a good indicator of the battery consumption. In cycleway networks located on mountain areas, rather than the distance between two points i and j one should consider the energy consumption (or even regeneration) occurring while riding from i to j which in principle is different from that from j to i.

We outline here the model for a generic cycleway network defined by a directed graph $G = (N, A)$, where each arc $(i, j) \in A$ is associated with a length d_{ij} and an energy consumption e_{ij}, such that, even if $d_{ij} = d_{ji}$, it may happen that $e_{ij} \neq e_{ji}$.

We assume that there exists a subset $S \subseteq N$ of candidate sites for the installation of a charging station (CS) and that c_i is the installation cost for each $i \in S$, with a total available budget F. Part of the input is a set P of itineraries; each $p \in P$ is defined as a sequence of m_p nodes $(i_1, i_2, \ldots, i_{m_p})$.

Now, let us consider a parametric maximum consumption δ, that is, the maximum amount of energy that can be used between two consecutive charging operations. From the given input we extract a set of stretches $R_\delta(P)$ obtained as sub-portions of the given paths in P, whose total energy consumption is greater than δ. Each of these stretches must contain at least one charging station to allow cyclists not to run out of charge during their trips.

We then define a stretch–node incidence matrix A such that, for every stretch $j \in R_\delta(P)$ and for every node $i \in S$, $A_{ji} = 1$ if node i belongs to stretch j, and $A_{ji} = 0$ otherwise.

An optimization model requires a binary variable x_i for every vertex $i \in S$ that is equal to 1 if a CS must be installed at i and 0 otherwise. This allows us to formulate the *Charging-Station Minimum Set-Covering Problem (CMMSP)* for general graphs as follows: for a given δ, at least one CS must be installed on at least one node of every stretch $j \in R_\delta(P)$.

$$\min \sum_{i \in S} c_i x_i \tag{16}$$

$$\text{s.t.} \sum_{i \in S} A_{ji} x_i \geq 1, \qquad \forall j \in R_\delta(P) \tag{17}$$

$$x_i \in \{0, 1\}, \qquad \forall i \in S. \tag{18}$$

The above problem is parametric in δ, which means that we should find the smallest value of δ such that the value of the objective function (minimum installation cost) is less than or equal to the budget F. The solution of the problem can be obtained through a binary search framework, like that presented for the linear cycleway case. In this case, however, instead of solving a simple path search problem, we have to solve a set covering problem, which is known to be very difficult to solve to optimality.

Note that, while the cardinality of $R_\delta(P)$ is polynomial in the number of paths in P, the number of paths in P can be very large and hence the same may hold for the main set of constraints.

6 Scenario Analysis

The models proposed above were applied to several case studies of cycleway networks having different shapes and sizes.

For each case study, the set of paths P has been generated so as to ensure *coverage* of most nodes in the graph using this procedure:

1. Select a set of origin/destination pairs located at opposite ends of the network by choosing several pairs with large Euclidean distances.
2. solve a *maximum flow* problem between each o/d pair where all arcs have capacity 1; this creates a flow of at least t if the graph is t-connected.
3. create o/d paths from the flow solution.

Let us briefly describe the considered cycleways or networks. For all our case studies, except the Asiago one, we have used a constant CS installation cost of $c_j = 2000e$.

The VenTo Cycleway

In this section we consider the entire extension of the cycleway, which has 172 nodes and 344 arcs. Given the structure of the network, we achieve almost total node coverage using only a few paths that stretch from the extreme nodes of the graph: four paths suffice to cover all nodes but two. For this cycleway, we have a budget of $b = 80{,}000$€.

The Asiago Plateau Cycle Trail Network

The Asiago Sette Comuni (seven towns) Plateau is located in the province of Vicenza, Italy. Stretching between Bassano del Grappa and the border with the Trentino-Alto Adige region, the network spans an area of about 50km by 40km, with the town of Asiago at its center. This instance has been taken from Belotti et al. (2024). The graph has 28 nodes and 88 arcs. We generated six paths for this instance and obtained a solution with a budget of 10,000€. The installation cost of CSs varies between 2000 and 4000€to account for the difficulty of installing a CS in remote areas.

The Tyrrhenian Cycling Route

The Tyrrhenian Cycling Route is an inter-regional cycle tourism route that connects the French-Italian border with the city of Rome along the Tyrrhenian coast. It covers a total length of approximately 870 km. The project is also part of the "National Tourist Cycle Routes" system. In addition to offering a coastal itinerary, the route passes through historic villages, protected natural areas, and sites of archaeological and environmental interest, aiming to promote both cycling and slow tourism along the Tyrrhenian coast. The graph is composed of 119 nodes and 236 arcs; only two paths are needed to cover the entire network. The budget considered in this case amounts to 40,000€.

The Sardinia Cycling Tourism Network

The Sardinian cycling tourism network is a regional project aimed at creating an integrated system of routes dedicated to cycling, tourism, recreation, and everyday mobility. The infrastructure is developed as a "loop" and connects coastal and inland areas of the island, also passing through areas of great environmental, historical, and landscape value. The main section, called Ciclovia della Sardegna, covers approximately 1147 km. The entire regional network covers 1916 km. The nodes of the graph are 98 with 404 arcs, and we generated 16 paths that reach more than 80% of the destinations. The budget allocated is 50,000€.

The Sole-Mare-Vento Cycling Tourism Network

The Sole-Mare-Vento Cycling Tourism Network integrates four major national cycle routes: the Trieste–Venice Cycling Route, the Adriatic Cycling Route, the VenTo route, and the Sun Route. Extending from the Slovenian border in the northeast to the Gargano promontory in the south. The network covers approximately 1950 km. The graph has 398 nodes and 1624 arcs. We generated 16 paths with a total budget of 40,000€.

The Flanders Tourist Leisure Cycling Network

The Flanders Tourist Leisure Cycling Network is a comprehensive cycling network that extends across the entire Flemish region of Belgium.[1] The network covers more than 14 thousand km. The network is modeled as a connected graph with 523 nodes and 4776 arcs. The budget is of 274,000€.

The characteristics of the case studies are summarized in Table 3.

From the computational point of view, the proposed solution approach showed to be effective and efficient. In all case studies, the optimal solution has been found. The computational times range from seconds (or less as in the case of Asiago) to more than one hour as in the challenging instance of Flanders. In the case of Flanders, the large computational time is due to the very large number of paths that have been

[1] https://metadata.vlaanderen.be/srv/api/records/c91e9b9d-6465-4dec-beeb-16fdc6d759a0.

Table 3 Characteristics of instances: dimensions and properties of the original graph

Instance	#nodes	#arcs	#paths	Budget (k€)	Extension (km)
VenTo cycleway	172	344	4	80	700
Asiago Plateau	28	44	6	10	300
Tyrrhenian	119	236	2	40	870
Sardinia	98	404	16	50	1916
Sole-Mare-VenTo	398	1624	16	40	1950
Flanders	523	4776	289	274	14,000

generated to account for the complexity of the network and the variety of itineraries. In Fig. 7 we can see the large number of o/d nodes (in green). In any case, considering the fact that the location of CSs is a strategic planning decision, a long computational time can be acceptable due to the long-term nature of the decisions.

Considering the tourist cycling networks analyzed, the optimal solution identified has practical implications that are useful for planning charging infrastructure both in large areas connecting different regions and in smaller territories. In all cases, the maximum distance between two consecutive charging stations obtained from the optimization model provides a detailed design indication that corresponds to the average distances typically covered by cycle tourists during a stage of their journey (Pileri et al. 2020). Moreover, once the locations for installing the charging infrastructure are identified, this result can generate added value by creating conditions for the development of small local economies that support the territory and prevent the abandonment of small settlements along tourist cycle routes, thus reducing the depopulation of more marginal areas. Cycle tourism can therefore become a driver for investment in these areas and foster the economic conditions that young families often seek in their daily lives, reshaping the future of these territories. With these results, it is therefore possible to integrate the planning of services for electric cycling mobility with a regenerative territorial development model, a topic still scarcely addressed in the literature (Pirolo and Moscarelli 2025) (Figs. 8 and 9; Table 4).

Fig. 7 Flanders region: yellow nodes represent the location of the CSs, green nodes are the endpoints of the paths. The distribution of CSs is uniform in the area outside the city of Brussels, which is located in the central gray area. The O/Ds are positioned at the extremes of the graph to achieve maximum node coverage. The longest stretch and indicates the trail arc from 124 to 53, with a length of 40.53 km

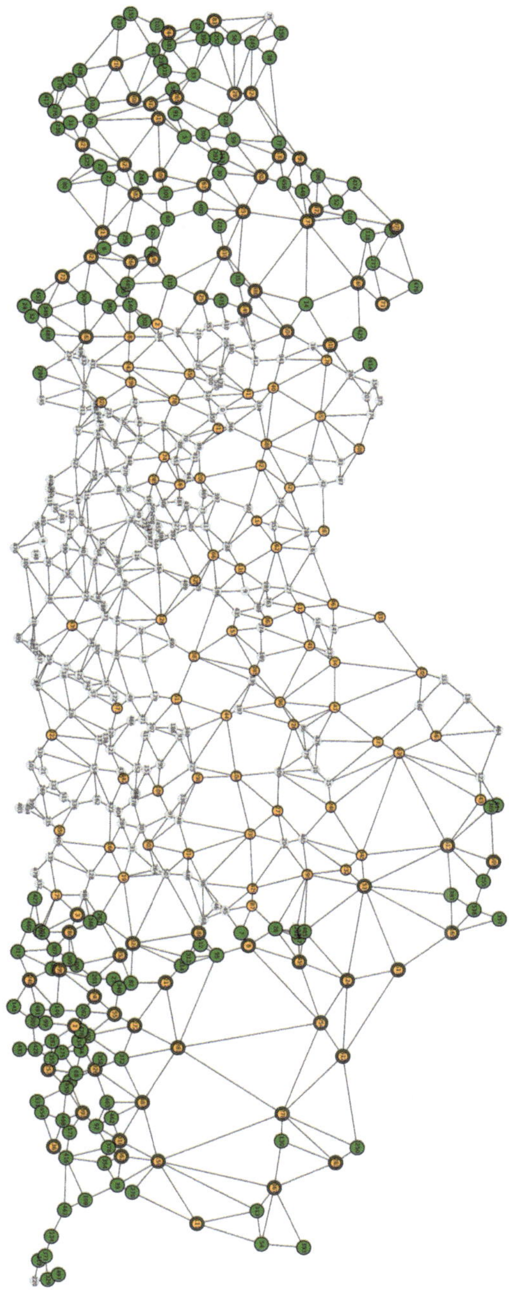

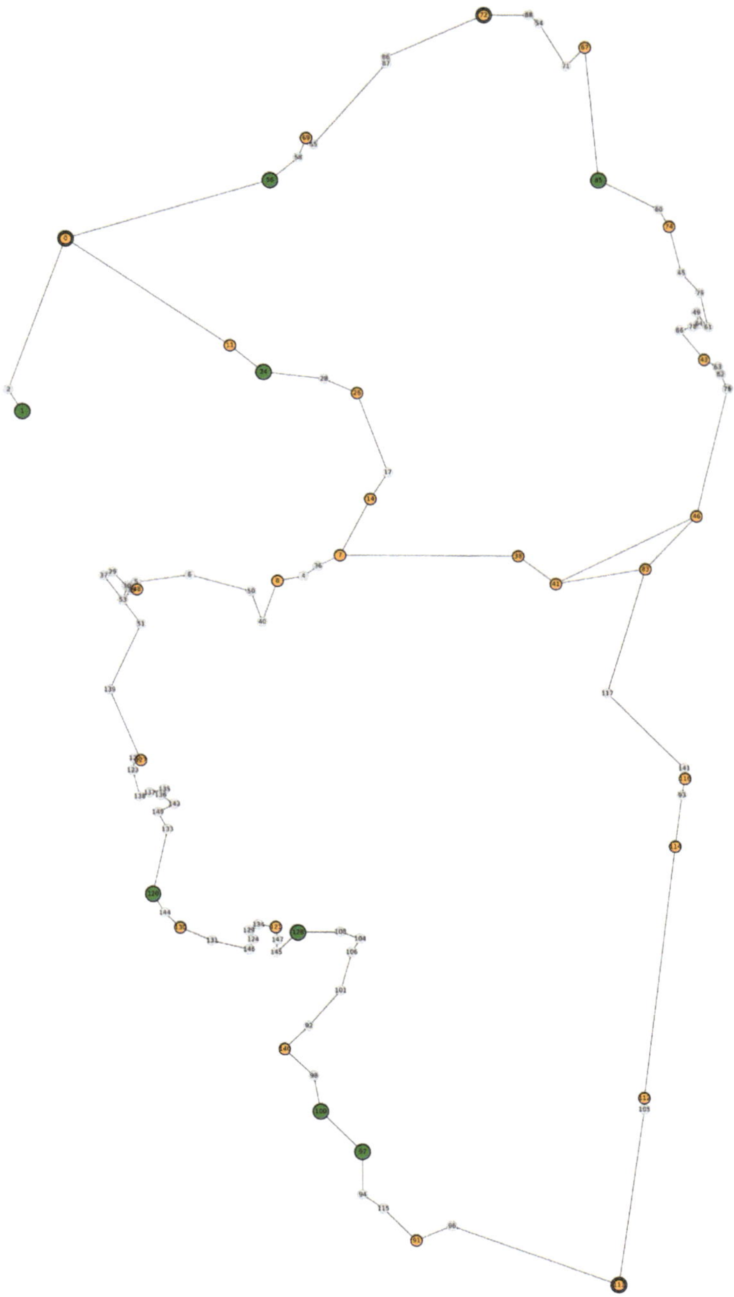

Fig. 8 Sardinia network: yellow nodes represent the location of the CSs, green nodes are the endpoints of the paths. The longest stretch is the trail arc from Bari Sardo to Villaputzu, with a length of 64.8 km. The origins or destinations of the paths are Porto Torres, Alghero, Ardara, Valledoria, Porto Pozzo, Olbia, San Sperate, Serramanna, Villasimius, Arborea, Baress

Fig. 9 Sole-Mare-Vento network: yellow nodes represent the location of the CSs, green nodes are the endpoints of the paths. The longest stretch is the trail arc from Termoli to Ortona, with a length of 62.4 km. The origins or destinations of the paths are Torino, Rozzano, Vieste, Trieste, Assago, Verona, Fiesole, Samone, Piacenza, San Giorgio di Nogaro, Albareto, Rivalta sul Mincio, Marzabotto, Greve in Chianti

Table 4 Solution times and optimal solution value

Instance	Total sol. time (s)	Max. distance (km)
VenTo cycleway	37.3	27.6
Asiago Plateau	0.1	17.1
Tyrrhenian	29.3	44.6
Sardinia	20.2	64.8
Sole-Mare-VenTo	1273.3	62.5
Flanders	4320.2	40.5

References

Belotti P, Errico F, Malucelli F, Massetti AT (2024) Optimization of e-bike networks. Transp Plan Technol 47(6):922–943. https://doi.org/10.1080/03081060.2024.2338873

Burani E, Cabri G, Leoncini M (2022) An algorithm to predict e-bike power consumption based on planned routes. Electronics 11(7):1105. https://doi.org/10.3390/electronics11071105

Caicedo A, Estrada M, Medina-Tapia M, Mayorga M (2023) Optimizing bike network design: a cost-effective methodology for heterogeneous travel demands using continuous approximation techniques. Transp Res Part A: Policy Practice 176 (Article 103826). https://doi.org/10.1016/j.tra.2023.103826

Černá ČA, Malucelli F, Nonato M, Polena L, Giovannini A (2014) Designing optimal routes for cycle-tourists. Transp Res Procedia 3:856–865. https://doi.org/10.1016/j.trpro.2014.10.067
Giovannini A, Malucelli F, Nonato M (2017) Cycle-tourist network design. Transp Res Procedia 22:154–163. https://doi.org/10.1016/j.trpro.2017.03.022
Hagberg AA, Schult DA, Swart PJ (2008) Exploring network structure, dynamics, and function using networkx. In: Varoquaux G, Vaught T, Millman J (eds) Proceedings of the 7th Python in science conference, SciPy2008, pp 11–15
Liu H, Szeto WY, Long J (2019) Bike network design problem with a path-size logit-based equilibrium constraint: formulation, global optimization, and matheuristic. Transp Res Part E Logist Transp Rev 127:284–307. https://doi.org/10.1016/j.tre.2019.05.010
Lonardi A, Szell M, De Bacco C (2025) Cohesive urban bicycle infrastructure design through optimal transport routing in multilayer networks. J Royal Soc Interface 22 (Article 20240532). https://doi.org/10.1098/rsif.2024.0532
Mauttone A, Mercadante G, Rabaza M (2017) Bicycle network design: model and solution algorithm. Transp Res Procedia 27:969–976
Ministero (2018) delle infrastrutture e dei trasporti. Decreto ministeriale numero 517. https://www.gazzettaufficiale.it/eli/id/2019/01/22/19A00326/sg
Natera Orozco LG, Battiston F, Iñiguez G, Szell M (2020) Data-driven strategies for optimal bicycle network growth. Roy Soc Open Sci 7(12):201130. https://doi.org/10.1098/rsos.201130
Paulsen M, Rich J (2024) Welfare optimal bicycle network expansions with induced demand. Transp Res Part B: Methodol 190 (Article 103095). https://doi.org/10.1016/j.trb.2024.103095
Pileri P (2020) Slow travel project for enhancing territories: motivations and directions. Cycling & walking for regional development. Springer, Cham, pp 129–140
Pileri P et al (2020) Progettare la lentezza. Linee antifragili per rigenerare l'Italia a piedi e in bici, People
Pirolo L, Moscarelli R (2025) Integrating cycling tourism, e-mobility and local development: a spatial model to plan slow tourist paths for e-bikes. Regional Sci Policy Practice 100237
Pirolo L, Moscarelli R (2026) Planning slow tourism with e-technologies. A proposal for locating e-bike charging stations along the Italian national system of tourist cycling routes. In: The future of slow mobility. A review on design, planning, and ethical aspects of e-bikes. Springer, forthcoming
Pirolo L, Belotti P, Malucelli F, Moscarelli R, Pileri P (2024) Optimal charging station location in a linear cycle path with deviations. In: International symposium on combinatorial optimization. Springer, pp 361–375
Steinacker C, Storch DM, Timme M, Schröder M (2022) Demand-driven design of bicycle infrastructure networks for improved urban bikeability. Nat Comput Sci. https://doi.org/10.1038/s43588-022-00318-w

MIX
Papier aus verantwortungsvollen Quellen
Paper from responsible sources
FSC® C105338

If you have any concerns about our products,
you can contact us on
ProductSafety@springernature.com

In case Publisher is established outside the EU,
the EU authorized representative is:
Springer Nature Customer Service Center GmbH
Europaplatz 3, 69115 Heidelberg, Germany

Printed by Libri Plureos GmbH
in Hamburg, Germany